The Good, the True, and the Beautiful

Published in association with Éditions Odile Jacob for the purpose of bringing new and innovative books to English-language readers. The goals of Éditions Odile Jacob are to improve our understanding of society, the discussions that shape it, and the scientific discoveries that alter its vision, and thus contribute to and enrich the current debate of ideas.

The Good, the True, & the Beautiful

A Neuronal Approach

JEAN-PIERRE CHANGEUX
translated and revised by Laurence Garey

Yale UNIVERSITY PRESS
New Haven & London

Éditions Odile Jacob
Paris

Translated from *Du vrai, du beau, du bien: Une nouvelle approche neuronale,* by Jean-Pierre Changeux,
published by Éditions Odile Jacob in 2008.

Yale University Press books may be purchased in quantity for educational, business, or promotional use. For information, please e-mail sales.press@yale.edu (U.S. office) or sales@yaleup.co.uk (U.K. office).

Set in Minion type by Integrated Publishing Solutions.
Printed in the United States of America by

Library of Congress Cataloging-in-Publication Data

Changeux, Jean-Pierre.
[Du vrai, du beau, du bien. English]
The good, the true, and the beautiful : a neuronal approach / Jean-Pierre Changeux ;
translated and revised by Laurence Garey.
p. cm.
Includes bibliographical references and index.
ISBN 978-0-300-16139-7 (hardback)
1. Neurosciences—Philosophy. 2. Neurophysiology—Philosophy.
3. Neuropsychology—Philosophy. I. Title.
QP356.C4713 2012
612.8—dc23 2011051135

A catalogue record for this book is available from the British Library.

This paper meets the requirements of ANSI/NISO Z39.48–1992 (Permanence of Paper).

10 9 8 7 6 5 4 3 2 1

Contents

Translator's Preface

In 1985 I translated Jean-Pierre Changeux's *L'homme neuronal* as *Neuronal Man,* and so I was very pleased when Jean-Pierre and Odile Jacob asked me to take on the English edition of *Du vrai, du beau, du bien.*

The original French edition, published in 2008, was compiled partly from lectures given by Jean-Pierre Changeux at the Collège de France over thirty years and was a natural progression from *Neuronal Man.* In this English edition, we took the opportunity to transform it into a series of reflections on the human brain from morphological, physiological, chemical, and genetic standpoints, as well as to put it clearly in the context of psychology, philosophy, and, perhaps above all, art.

The title is based on Plato's definition of the universal questions of the natural world. He saw the Good, the True, and the Beautiful as independent celestial essences, but so intertwined as to be inseparable. This book uses a top-down approach to place the Good, the True, and the Beautiful within the human brain's neuronal circuitry.

Molecular biology and genetics of the brain are described in terms of molecules and the mind, notably neurotransmitters, including acetylcholine and its receptor.

More global consequences are introduced, such as neuroesthetics: the definition of beauty and how we perceive it, artistic creation in plastic arts and music, and the physiology of collecting and collectors.

Consideration is given to ethics and morality and their evolution through genetic and epigenetic factors, as well as to consciousness, learning, language, and writing. These more moral and social aspects of the brain are related to the theory of a conscious neuronal workspace as a reality within the brain circuitry. Illustrations are derived from the phenomena of waking and sleeping, anesthesia, and consciousness of death.

I want to thank Jean-Pierre for thinking of me in the first place, Odile Jacob for her support and encouragement, and Jean Thomson Black of Yale University Press for her tireless attention to the numerous pitfalls unavoidable in such an enterprise.

Laurence Garey

Acknowledgments for the French Edition

This book owes its existence to the inspiration and extraordinary competence of Odile Jacob, her constant striving for excellence, her energy and intelligence, which have all allowed it to be produced in a very short time.

It also owes much to the application and critical judgment of Claude Debru, who contributed directly to the choice and logic of the presentation of the texts, as well as to the editorial expertise of Marie-Lorraine Colas.

These texts bear witness to thirty years of teaching at the Collège de France, and I thank its administrator, Pierre Corval, for authorizing their publication.

JEAN-PIERRE CHANGEUX

The Good, the True, and the Beautiful

Introduction

The good is the beautiful . . . and the beautiful is the good.

—Plato, *Dialogues*

In Defense of Neuroscience

The human brain is the most complex physical object in the living world. It remains one of the most difficult to understand. If you tackle it head-on, you risk total failure. In the jungle of nerve cells (the neurons) and their interconnections at synapses, of which it is composed, we must try to identify pertinent features of its organization and function, Ariadne's thread to the center of the labyrinth. My thirty years of teaching at the *Collège de France* have provided me with an exceptional laboratory of ideas to help grasp this thread. They have had a profound influence on my theoretical reflections, too often bridled by the empirical constraints of the other laboratory, that of pure science. They have afforded me wonderful freedom, limited only by the severe criticism of a faithful but exacting audience.

My first seven years were summed up in *Neuronal Man,* first published in English in 1985. In the chapter titled "Mental Objects" I tackled an entirely new theme for me, that of the higher functions of the brain and consciousness. Later I seized the opportunity to analyze much more deeply those cognitive functions to which the electric fish, the mouse, or the rat, with which we usually work, do not give us direct access. If we wish to think effectively and make progress in our understanding of the brain, we must first take into account the multiple, parallel hierarchical levels that underlie its functions. Otherwise, we run the risk of taking the human brain as an all too simple collection of genes, neurons, and microcircuits, or of reviving a totally obsolete dualism.

In the last few decades sciences related to the nervous system have totally changed in outlook. It is no longer a habit, as was once the case, to make one's mark in a narrow discipline, entrenched in one's own physiological, pharmacological, anatomical, or behavioral culture. With molecular biology on the one hand and cognitive science on the other, a *nuovo cimento,* new conceptual as well as methodological syntheses, has become possible, uniting various approaches to the nervous system. During the 1980s molecular genetics prepared the ground for large-scale sequencing of genomes, providing a mass of new data with wide applications, especially in the domains of physiology, pharmacology, and pathology. With the perfection of methods for imaging the living human brain, another discipline, that of physics, opened the way to novel investigations of the relationship between mental state and physical activity in the brain. The common striving for conceptualization and theoretical modeling within these disciplines fertilized a new field of research, neuroscience, born in 1971 with the first meeting of the Society for Neuroscience in the United States. Even if the neuroscience revolution has happened, it has not yet borne all its fruit; far from it. We must now cross, step by step, the immense terra incognita that still separates the biological sciences from the science of man and society.

So, we shall take a neurobiological approach to our discussion of three of the universal questions of the natural world, as defined by Plato (428–348 BCE), and by Socrates (469–399 BCE) through him, in his *Dialogues.* He saw the Good, the True, and the Beautiful as independent, celestial essences or Ideas, but so intertwined as to be inseparable. The Good was the True, which was the Beautiful. If we saw any conflict among them it was because we failed to perceive them perfectly: the Good, the True, and the Beautiful, being the same, could not be in conflict. Thus, if Good is beautiful and Truth is beautiful, let us begin by talking of beauty. This leads us to a top-down approach, contrary to Plato's, which we shall maintain throughout the book to place the Good, the True, and the Beautiful within the characteristic features of the human brain's neuronal organization.

One of the major unsolved questions of today's neuroscience is what it means to "perceive" something and how we are aware of our own existence. Although we know quite a lot, and more every day, about the wiring and chemistry of the brain, there is still an immense gap between that knowledge and turning molecular biology into consciousness. Scientists will be trying to bridge this chasm for a long time to come. This book does not claim to do that in any unified way. It is, as I have said, a presentation of ideas based on academic lectures and personal experience, a series of experimental forays into the particular aspects of the central nervous system that are amenable to direct investigation. There will be gaps and discontinuities because of the way the topics were approached in the first place.

I
The Beautiful
NEUROESTHETICS

What Is Neuroesthetics?

The term *neuroesthetics* is of recent origin. It was coined by Semir Zeki, and the first conference on the theme was held in San Francisco in 2002. It reflected a somewhat older concept, such as that expounded by Alexander Luria in the 1970s, aimed at finding the neural basis for contemplating and creating artworks and studying it scientifically. In the next pages I shall attempt to link some personal aspects of art and esthetics to various biological observations, in the hope that the reader will accept plausible, but not definitive, interrelationships.

Ethics strives for a life of "goodness"; science seeks universal "truth" and implies cumulative acquisition of knowledge, whereas art seeks interpersonal communication of "beauty." It involves motivation and emotion in harmony with reason, without obvious progress but constantly renewed. Of Plato's three principal fields of human culture—ethics, science, and art—I believe the last to be the oldest, for it has been found among nonhuman animals. It is basic to reinforcing social ties by reason of the universality of the forms of communication it utilizes. From my point of view, cognitive function, especially consciousness and artistic activity, developed in parallel with a major reorganization of the brain, notably by the expansion of the cerebral cortex and especially the association cortex of the prefrontal, parietotemporal, and cingulate areas, in close relationship with the limbic system. This association cortex neither receives direct sensory input nor influences motor activity directly, but it is connected to other cortical areas. The limbic system includes those parts of the brain, both cortical and subcortical, that deal with, among other things, emotion, drives, and motivation.

Several ideological assumptions, which are common in the human sciences, must be dismantled. The first problem is dualism as part of the body-mind problem. Modern neuroscience is removing this archaic distinction, which was based on the deliberate disregard of scientific progress. Neuroscience, indeed, has established reciprocal causal relationships between neural organization and the activity it generates, which can be seen in the progressive development of behavior or mental processes. The extreme complexity of the functional organization of our brain, mainly unsuspected as it was, must now be reckoned with. It involves multiple forms of evolution, past and present, each nested one within another. They are genetic and epigenetic (what comes above and beyond genetic constraints, as we shall see in detail later), developmental, cognitive, mental, and sociocultural, each leaving a singular material track in this organization.

The second problem is that of nature and culture. It should not be confused with the much better-defined question of nature and nurture (the innate and the acquired), which differentiates between what is determined genetically and what is learned later. Understanding the innate requires the elucidation of the still poorly understood relationship between the human genome and the phenotype of the brain. Understanding the acquired demands an analysis of epigenetic control of synaptic development, that is, the connections within the brain, including spontaneous activity in the nervous system in addition to activity provoked by interaction with the environment. Culture depends on epigenetic plasticity in developing nerve networks. Paradoxically, we can say that culture is first and foremost a biological, or rather neurobiological, track. So there is no conflict between nature and culture. On the contrary, the genetic envelope, the constraints imposed on the human brain by heredity, includes epigenetic response to the environment and, therefore, genesis of culture. The singularity of the history of human populations, and their individual histories, will therefore materialize in the form of neural tracks that I call *neurohistorical objects,* without which history does not exist. Nonhuman animals have no history, except at the level of their genes; human beings have a history in their neural organization nested in their genome.

The last problem is that of the spiritual and the material. In everyday language there exist often deliberate confusions in meaning, derived from a supposedly dualist ideology still rampant today. The spirituality of intellectual, philosophical, religious, and esthetic activities is contrasted with the baser needs of everyday life, such as survival and primitive pleasures. Denis Diderot (1713–1784) in his *Elements of Physiology,* which he, sadly, never finished, attempted to abolish this distinction by showing that the highest human "spiritual" activities were in reality a manifestation of the organization of our brain; his intention was in no way to diminish the quality of this secularized spirituality, but to emphasize its dignity. Such a concept opens a vast horizon of human science to a neurohistorical approach.

How Can We Define Beauty?

What is beauty? How can we distinguish a "beautiful" work of art from one that is ordinary or even ugly? Diderot noted that most of humankind agreed that beauty existed, but though they recognized it, very few knew what it was. The ancient Greek philosophers considered art as imitation, *mimesis,* or reproduction of reality. Plato's mimesis was to copy nature, like trompe l'oeil, which in itself posed the philosophical problem that reality was already a copy of the world of the intellect. So the artist's work was a copy of a copy. It was thus not only useless, for it duplicated reality, but dangerous, for it was deceitful. Plato's *Republic* provided an interesting clarification. He took a bed as an example. He distinguished the idea of a bed at the intellectual level from a designer's model, then from the object itself as made by the carpenter, then from a painted picture of a bed. In neuroesthetic terms, we recognize a concept, an object, and the conceptual picture of an object. More recently, we might recall René Magritte's 1929 painting of a pipe, which he distinguished from reality by titling it *Ceci n'est pas une pipe* (This is not a pipe).

For Aristotle (384–322 BCE) art was a human activity like any other: we must understand it, and its "causes," before condemning it. For example, a statue was made of marble (the *material cause*) by a sculptor (the *efficient cause*) according to a pattern (the *formal cause*) with a certain aim in mind (the *final cause*). Imitation extended nature and was praiseworthy. If we liked pictures, we learned from looking at them. Art was a natural tendency, a source of pleasure, a source of learning. Mimesis was not deceptive: we could always distinguish the real from its image. We were not stupid like Zeuxis' birds pecking at painted grapes. The representation of the ugly could be beautiful. The onlooker could rid himself of his emotions at secondhand: catharsis. So art was not only pleasant, but useful. It was not simply a copy of nature but an idealized "representation" in its own right. Georg Hegel (1770–1831) took up the theme of mimesis. For him simple imitation was a caricature of life. Art was much more, a union of sensation and intelligence, created by the sprit, the brain.

Paradoxically, the artist must sometimes "cheat" to be realistic. One example is the illusion of galloping horses traditionally painted with all four hooves off the ground, as in Théodore Géricault's *Derby at Epsom* (Figure 1). Auguste Rodin (1840–1917) saw a painting as more truthful than a photograph, "for in reality time does not stop." In Impressionism the juxtaposition of pure colors introduced a new code of realism. With abstract artists, such as Wassily Kandinsky (1866–1944), the observer has to learn to see paintings as representing a state of mind, not an actual object. According to Piet Mondrian (1872–1944), in abstract art man attains a much deeper vision of sensory reality. For Paul Klee (1879–1940) art does not reproduce what is visible; it makes it visible, changes our way of looking at things, and teaches us to see. It is a deliberate attempt to represent transitory, personal images from the conscious realm of an artist in a stable, public medium. In surrealism and hyperrealism a new mimesis developed, inverting trompe l'oeil. Ready-made objects, such as newspapers and labels, appeared in the picture. Art went from figuration to transfiguration of reality.

For the neuroscientist the question of mimesis is linked to

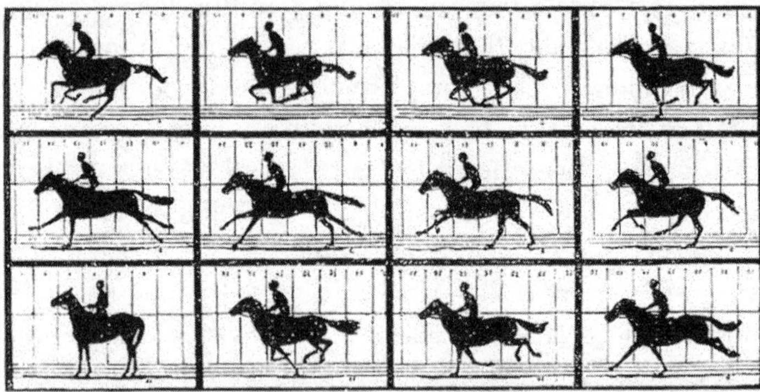

Figure 1: *Mimesis in art.* Comparison of Géricault's *Derby at Epsom* (1821) with Eadweard Muybridge's high-speed photographs from his work *Animal Locomotion* (1887), showing the difference between an artist's impression of a galloping horse and a scientific observation. Géricault did not fully respect reality but simulated the movement.

that of visual perception. To what extent is sensation "bottom up" (or instructive) or "top down" from conscious perception? What is the importance of detachment or mimesis in re-creating an ideal view of nature? How does intelligence interact with sensation in the inner world of conscious esthetic space? Are Kandinsky's states of mind understandable in terms of endogenous spontaneous activity, or even hallucinations? What did Kandinsky mean by impression, improvisation, or composition? By juggling with his "codes," Leonardo da Vinci (1452–1519) revealed that the artist painted himself. So is the human brain a reflection of society, as Karl Marx suggested, or does society reflect the brain? All in all, we can say that a work of art shares much in common with a scientific model. It is both reductive and revealing: it aims to be communicated socially and to be received and shared by a group.

Let us return to the definition of beauty. Plato's *Hippias major* provided an important clarification. Beauty was manifest thanks to formal appropriateness, where the unity of the whole triumphed over the multiplicity of its parts. Since ancient Greece mathematics has provided an excellent means of mastering the relationships between the whole and its parts. So there is the mathematics of appropriateness. Its expression takes various forms, including repetition and symmetry. The Greeks saw *symmetria* as comparing two elements. Plato's regular solids (such as pyramids, cubes, and so on) bear witness to a harmony imposed by the demiurge on the primal chaos of the universe. Was beauty the optimal form of fitness for purpose? Plato's reply in *Hippias major* was affirmative, and he recognized a beauty of usefulness. Socrates advocated that beauty was both useful and appropriate, a product of excellence. We see it today in technology and design. An object is beautiful if it is fit for its purpose, the very principle of simplicity and economy. For Leon Alberti (1404–1472) and the Stoics beauty was again the harmony of all parts in relation to one another, and for René Descartes (1596–1650) it was the perfect agreement of all parts together. Diderot said, "All unity is born of the subordination of the parts, and from this subordination is born harmony within variety." He cited Anthony Ashley-Cooper, First Lord Shaftesbury (1621–1683), who said that a beautiful man was one whose well-proportioned body best suited him to accomplish his animal functions. This fits well Charles Darwin's (1809–1882) concept of sexual selection. In Plato's *Symposium* beauty is defined on the basis of quality of amorous activity. Indeed, sexual references are frequent in definitions of beauty.

Defining beauty raises important physiological questions. Can the perception of relationships in a painting be compared to intrinsic harmonic rhythms in music? Does the perception of the coherence of the parts with the whole relate to that of space and its ego- and allocentric components? If the universe of art is a recomposed universe, is there some final purpose? Whatever the case, it is difficult to propose a general definition of beauty to which there are no exceptions, perhaps as difficult as Georges Canguilhem (1904–1995) felt it was to define life. Perhaps we need only select certain significant features without attempting a single, restrictive definition.

From the Light of Antiquity to Today: An Introduction to Plastic Arts

In Aristotle's *Metaphysics* vision is the sense that enables us to acquire most knowledge. The ancient Greeks showed great interest in the relationship between the eye and an object. The doctrine was that there were visual beams and that a fire shone out from the eye under the influence of light and contacted objects to determine their shape, color, and other properties. Euclidean optical geometry of about 300 BCE postulated a cone with the eye at its apex, a theory not unlike modern concepts of visual receptive fields—that is to say, that part of our visual field which stimulates a given cell in the visual pathways of the brain. Another doctrine was that of Epicurus (341–270 BCE), in which particles or fine films left an object and entered the eye, thereby transmitting its shape. Aristotle proposed a transparent medium between the observer and an object that changed qualitatively during vision, transmitting the form of the object but not its matter. The sensation was real, and he pursued a remarkable psychological analysis: common sense distinguishes various perceptions and what they have in common, whereas *imagination* retains images of what the sense had perceived. Finally, memory represents the image as belonging to the past and recognizes similarity between two successive sensations. Memory makes possible experience, the source of all authentic science. Aristotle went beyond a purely empirical stance when he wrote that science is not possible through sensation alone: one has to extract the specific features of sensory objects.

In the tenth century Ibn al-Haytham (Alhazen) made a major advance in the understanding of the propagation and perception of light. He proposed, rightly, that light rays are propagated from an object to the eye, which is a light receptor. He was wrong, however, in suggesting that it is the lens, not the retina, that constructs a point-by-point image and transmits it via the optic nerve to the "seat of the soul." Indeed, it was the astronomer Johannes Kepler in 1604 who finally recognized the role of the retina in visual perception. Alhazen completed his *Book of Optics,* similar to Euclid's work but with opposite conclusions, by distinguishing the aspect of a sensation from its intuition or knowledge, which implies the intervention of perceptive judgment.

Two early "artificial" methods should be mentioned here. First, the portrayal of reality through perspective attempted to give the illusion of solid construction and broke with the medieval concept of art representing an inner idea for which symbolic features sufficed. Artists of the Trecento, including Duccio and Giotto, adopted the tradition of illusion from the Greco-Romans, such as those at Paestum and Pompeii. Using perspective, they attempted to simulate natural vision in their works. Second, the development by craftsmen of optical instruments such as the telescope, which Galileo turned toward the heavens, meant that our knowledge of the world was no longer restricted by the limits of our natural senses. Humanity's vision extended "much further than the imagination of our fathers was wont to go" (Descartes, 1637).

Johannes Kepler (1571–1630) was concerned with errors of vision, especially in relation to astronomy. In *Astronomia pars optica* (1604) and *Dioptricae* (1611) he proposed a mathematical theory of the camera obscura, which was based on light rays penetrating a tiny opening and projecting a reversed image on a white screen. He extended this model to the eye, whereby the pupil is the opening and the retina is the screen. There would be a two-dimensional image of the object, a physical entity in itself that the observer would "see" directly. This idea was adopted and even illustrated by Descartes in his *Dioptrics* (Figure 2). In the same work he proposed a rational theory of the telescope and therefore of vision, from object to eye, and from eye to the "innermost folds of the brain," read by the "soul" via the pineal gland.

Isaac Newton (1642–1727) questioned the nature of light

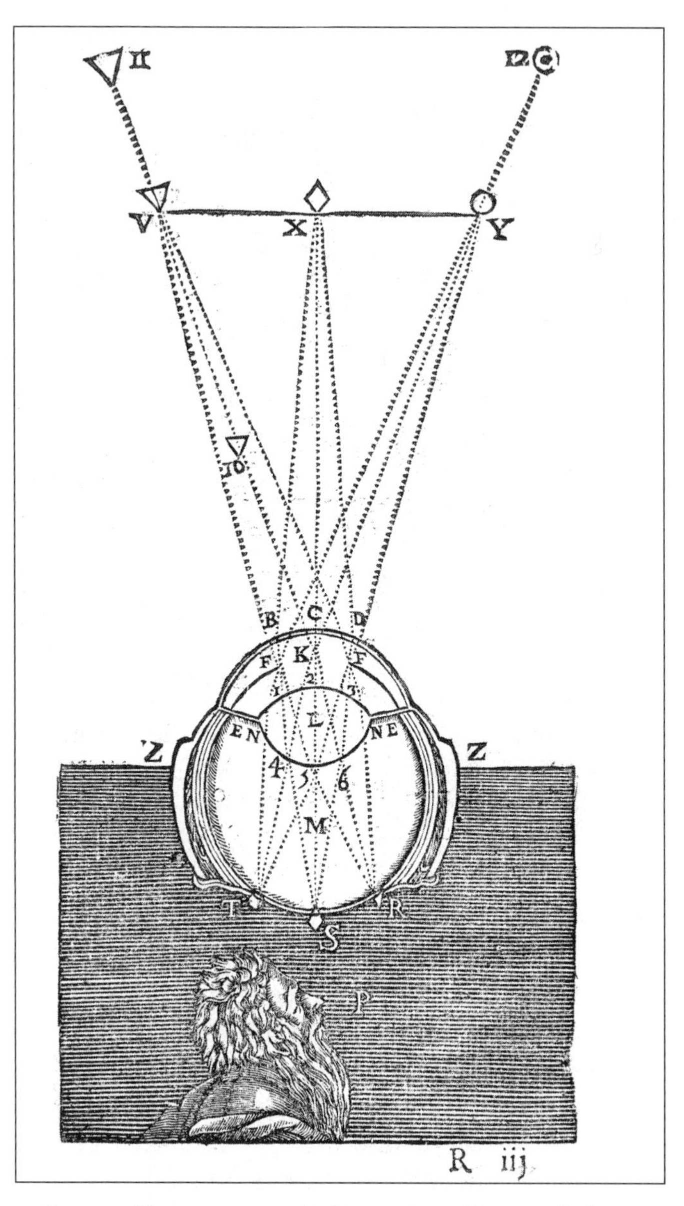

Figure 2: *The eye as an optical instrument.* Descartes's figure from his *Dioptrics* (1637) of the projection of the outside world on the retina. It obeys the laws of geometrical optics. Collection of J.-P. Changeux.

itself. Is white light pure and homogeneous? Do colors derive from a weakening of incident light (that is, direct light that falls on a surface) by mixing with darkness, as was supposed since Aristotle? Newton's experiments ran in the following sequence. In his first experiment of 1665 he observed a ribbon of two colors, half blue, half red, through a prism, and he saw that the blue was closer to the angle of the prism than the red. Some time later he described a parallel incident ray of light, coming from a crack in the shutters of a dark room, on a prism. The refracted rays did not produce a circle on the wall opposite but an oblong strip with blue and red at its edges. These two experiments demonstrated a specific refraction of different light rays. Next came a crucial experiment, described in a letter to Henry Oldenburg, the secretary of the Royal Society. Newton used a second prism to study individual refracted rays coming from the first prism. He showed that the rays crossing the second prism preserved their color and their degree of refraction. He concluded that colors are innate properties of different light rays, and that their mixing produces apparent transformations of color. Finally, he placed a convergent lens in front of the whole spectrum of colored rays coming from a prism and observed that the resultant color was completely white. He proposed that white light is a complex mixture of rays of different colors emitted indiscriminately from luminous bodies. There exists an orderly succession of colors, as seen in a rainbow, from the least refracted, bright red, to the most refracted, violet.

From then on, two theories of the nature of light were in competition. Robert Hooke (1635–1703) compared a light ray to a vibrating cord, whereas Newton held that a luminous body emits minute corpuscles or particles. In 1905 Albert Einstein (1879–1955) reconciled the wave and particle theories of light by introducing the photon—a quantum of light energy transmitted like a wave. For James Clerk Maxwell (1831–1879) light was part of a series of electromagnetic radiations, each with a wavelength of 370 to 730 nanometers, produced by the movement of charged particles, such as electrons, or by changes in energy levels. So the spectrum of wavelengths of daylight is different from the much more selective and colder one of a fluorescent lamp. The color of an object is the result of absorbed wavelengths compared with reflected ones. For example, a red area in a painting absorbs short wavelengths (blue and green) and reflects the longer reds.

Newton's theories were accepted by the European Enlightenment and by modern science. They implied the application of experimental method as outlined by Francis Bacon (1561–1626), the reduction of overall complexity into simpler elements, and the mathematization of physics (proposed by Galileo), followed by that of the universe (Pierre-Simon Laplace, 1749–1827), and natural and social phenomena (Pierre de Maupertuis, 1698–1759). Scientific theory opened a separate and autonomous field within which one could propose an explanation of the world without reference to a divinity, even if Newton was not only religious, but an alchemist, kabbalist, and millenarian. The separation of science and religion conferred a new dignity to science, which was always conscious of its limitations.

The Eye and Light Receptors

The vertebrate eye is a remarkable optical instrument in which a lens focuses a visual image on the retina. The sensitive layer of the retina is formed of rod and cone receptor cells, which relay to a very complex neuronal network, the output elements of which are the retinal ganglion cells, from which up to 1.5 million outgoing nerve fibers (axons) form the optic nerve. The rods respond to relatively low intensities of light, but without color information, whereas cones are color-sensitive. Thomas Young (1773–1829), in his Bakerian Lecture of 1801 on color perception, proposed for the first time the existence of three categories of receptive elements in the retina that are tuned to wavelength. Each sort of cone contains one set of photoreceptive molecules specific for red, green, or blue. In fact, there are four photoreceptive pigment molecules: *rhodopsin* in the rods, which have a wider spectrum, and the three *cone opsins* (Figure 3). They are transmembrane allosteric proteins, each with a distinct absorption spectrum, and are bound to a retinal chromophore, a form of vitamin A. The absorption maximums for each visual pigment are about 570 nanometers for red, 540 for green, and 430 for blue. These proteins have seven transmembrane domains coupled to so-called G proteins, with considerable homologies between them, and only a few different amino acids. Glutamate and tyrosine on domains 2 and 5 are responsible for spectral tuning. The genes coding for rhodopsin and the opsins have been identified and sequenced. Hereditary changes in these genes lead to color blindness. Color perception, like that of shape, is determined at the retina by strict molecular mechanisms.

In addition to color, artists speak of tone. In scientific terms this means the luminance of a painting. Is it lighter or darker? Luminance refers to the perceived light radiated by a source in relation to its wavelength. The concept plays a critical role in the perception of depth, movement, and spatial orientation. Cones behave differentially in luminance discrimination: the response to green light is twenty times greater than that to blue. The final response to luminance is, in combina-

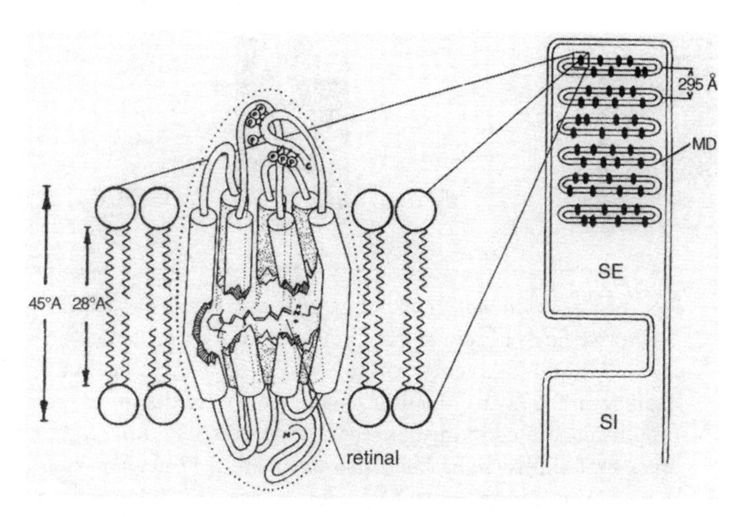

Figure 3: *The visual pigment rhodopsin.* At its center is the retinal molecule. The right-hand drawing shows its localization in the external segment (SE) of a retinal rod. From P. Buser and M. Imbert, *Neurophysiologie fonctionnelle.*

tion with rod activity, by the ganglion cells. Night vision is essentially due to rods alone and is therefore color-blind. Rods are in fact more sensitive to green and blue, and they dominate the total effect of the cones. Reds seem darker and blues lighter, a phenomenon dubbed the Purkinje shift. So there is no absolute perception of color, but rather a reconstruction of color, like that of the whole outside world, by the brain.

Antagonist Cells, Concentric Fields, and Shape

The first processing of visual information is by the retinal cells. In the 1950s Stephen Kuffler defined a physiological paradigm that is still important today, the receptive field. He stimulated the retina with a moving spot of light while recording from a

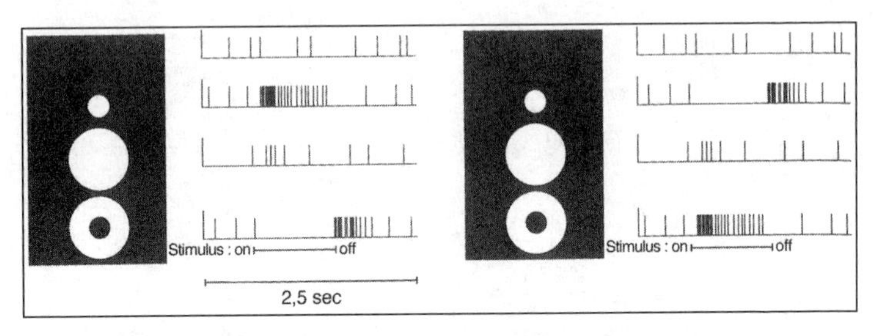

Figure 4: *Activation of "center-surround" neurons in the retina.* Receptive fields of retinal ganglion cells. Left: ON center, OFF surround. Right: OFF center, ON surround. The neuronal activity (spikes on the horizontal time axis) is shown for the specific stimulus illustrated in the black bars (from top to bottom: light center spot, diffuse light, light surround). From *Vision and Art: The Biology of Seeing* © 2002 Margaret Livingstone, Published by Harry N. Abrams, Inc. All Rights Reserved

ganglion cell. The part of the retinal surface producing a response in the cell was its receptive field. A small circular spot typically gave an "ON" response (indicating rapid electrical discharges when the light went on), but when the spot was made larger, the cell stopped firing. If this receptive field was stimulated by an annulus, a ring with a dark center, the cell responded when the stimulus was turned off. This cell had a concentric field that was ON center, OFF surround, but others had the opposite organization. Careful analysis of ganglion cell responses showed that they responded best to light/dark borders (Figure 4). In Ludimar Hermann's grid illusion of 1870 (Figure 5), gray spots appear at the intersections of a white grid between black squares: this phenomenon was interpreted as differential inhibition of antagonist cells with concentric fields by four white segments. Antagonist cells generally respond to abrupt changes in light and dark. They contribute mainly to recognition of the form and outline of a shape, much

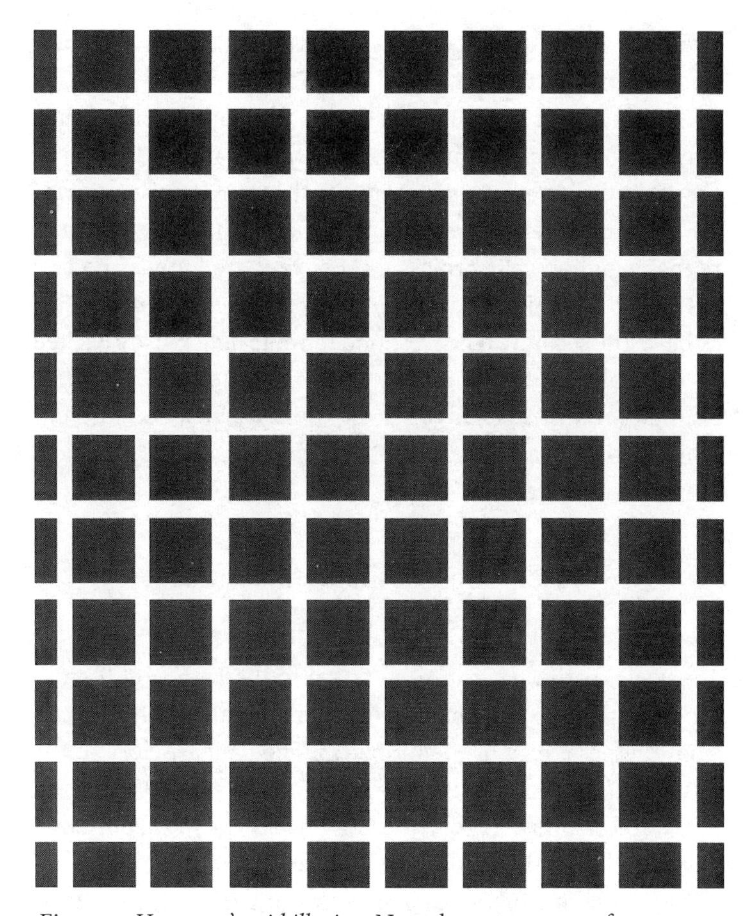

Figure 5: *Hermann's grid illusion.* Note the appearance of gray spots at the intersection of the white lines. This illusion is due to differential activation of the center-surround neurons of the retina. From *Vision and Art: The Biology of Seeing* © 2002 Margaret Livingstone, Published by Harry N. Abrams, Inc. All Rights Reserved

like what an artist paints on his canvas, even if the traced outline does not exist in nature.

Concentric antagonist cells are also encountered in the lateral geniculate nucleus (LGN) of the thalamus, the part of the brain that relays sensory input to the primary visual cortex for final interpretation as a sensation, to which the axons of retinal ganglion cells project. Three types are found. The first have small receptive fields with centers responsive to red, green, or blue. The second do not respond to white light but are excited by one color and inhibited by the complimentary color (for example, red ON, green OFF). The third participate differentially in the so-called dorsal "where" stream and are colorblind, or in the ventral "what" stream and are color-sensitive.

Preservation of the Retinal Image on the Cerebral Cortex

The visual pathways, from the retina to the LGN and then to the visual cortex, maintain an accurate topography. Axons from the nasal part of the retina (the medial part close to the nose) cross the midline, whereas temporal (lateral) ones do not. Cerebral imagery, whether PET scans (positron emission tomography) with a radioactive tracer such as deoxyglucose, or functional magnetic resonance imaging (fMRI), enables us to trace a distinct image of the stimulus, but with mathematical transformations analogous to those suggested by D'Arcy Thompson in 1917. This image is relayed from map to map in the various functional levels of the cortex, from primary to secondary and higher visual cortical areas as far as the prefrontal lobe, as first demonstrated by Roger Tootell and colleagues in 1982. So, as is consistent with an empirical concept, there is an ascending isomorphism of the representation of the outside world, which is accompanied by what one might call a descending egomorphism from the top down. It is less figurative, and its code remains to be determined. Tootell later used high-resolution fMRI in human volunteers who were stimulated visually with checkerboard images or dilating and contracting circles. This revealed maps whose contours varied significantly, but, importantly, in spite of this anatomophysiological variability, the perceived image was constant.

In the 1950s David Hubel and Torsten Wiesel founded physiological research on primary visual cortical neurons by recording single cells responding to stimuli in the visual field. They were unable to record responses using the sorts of spot stimuli that worked well in the retina, the optimal responses being to a moving bar of light oriented in a specific direction. They distinguished several cell types with various specific features, such as speed of movement and direction, as well as orientation. These cells were organized in vertical columns, through the cortex, of orientation specificity or dominance by one eye or the other.

A complete lesion of the primary visual cortex of one hemisphere can lead to hemianopia, blindness in the visual field on the opposite side. An incomplete lesion may cause a scotoma, a partial loss in the visual field. Lesions in other parts of the cortex may cause loss of color or movement perception.

Parallel and Hierarchical Organization of the Visual Pathways

The number of cortical representations of the retina has increased during evolution in parallel with the relative size of the cerebral cortex, especially the frontal cortex. In lower mammals there are three or four cortical visual areas, fifteen to twenty in lower primates, some thirty-two in the macaque, and probably many more in humans. David Van Essen has determined over three hundred reciprocal connections between the visual areas of the macaque, comprising parallel pathways and at least ten hierarchical levels. In his seminal work in the 1970s, Semir Zeki demonstrated that these multiple areas were organized specifically for perception of, for example, orientation, color (visual area V4), or motion (V5). In fact, this specificity was not absolute, and there was overlap between areas. In 1982 Leslie Ungerleider and Mortimer Mishkin proposed an alternative way of defining visual pathways. It uses the distinction between the small-celled (parvocellular) layers of the LGN (P pathway) and the larger-celled magnocellular layers (M pathway), described in detail by Margaret Livingstone and David Hubel in 1984. The P pathway projects to the primary visual cortex (V1) and thence to V4 and the temporal cortex (the ventral stream), and it deals with recognition of objects, faces, and color: it is the *what* pathway. The M path is through V1 to V5 and the parietal cortex (dorsal stream) and deals more with movement, separation of objects from background, and the organization of the visual scene in space: it is the *where* pathway. Certain artists seem to privilege one or other of these streams (Mondrian the P, Pol Bury the M). In 1981 Isia Leviant created *Enigma,* an illusion of movement from fixed concentric circles that perhaps is due to interaction between the two streams.

Color Vision

In 1860 Hermann Helmholtz adopted Thomas Young's theory of 1801, mentioned above, in his three categories of fibers, but he introduced a new idea: that each fiber type responds to several different wavelengths, but with a specific preferred frequency. There were several problems with Newton's color theory. The first was the interrelationship of colors themselves. For Newton there was continuity within the spectrum, as the passage from one color to the next proceeds according to a "color circle." This concept was opposed by the antagonist colors of Goethe (1810) and Hering (1874): the four colors red, green, yellow, and blue were considered primary. The antagonist pairs such as red and green or yellow and blue were complementary colors and canceled each other to produce a neutral color, black, white, or gray. The theory predicts the perception of millions of colors on the basis of only three receptor types. A computer screen can produce 17 million distinct colors. From Aristotle to Goethe color was supposed to be a mixture of black and white, dark and light. What a historic error!

The enlightened Jesuit Louis Bertrand Castel (1688–1757), a friend of Jean-Philippe Rameau, oddly opposed Newton's theories, not accepting the continuity of the spectrum, but distinguishing hue from tone, or chiaroscuro. He took up an old but important question: the harmonic relationships between sounds and colors. Giosoffo Zarlino (1517–1590) proposed a harmonic theory for sounds, the tetrachord (1+2+3+4=10), in which the ratios give the octave, the fourth, and the fifth. In 1613 François d'Aguilon suggested a concordance between sounds and colors: the relationships between primary and secondary colors and their origin from black and white resembled those of sounds. In 1725 Castel proposed an ocular harpsichord, which played sequences of colors like sequences of notes. He suggested an octave of twelve colors corresponding to the scale plus four half tones. The instrument was made in 1754, was as tall as a man, and contained sixty pieces of colored glass and five hundred lamps. When a key was pressed, a lamp appeared behind the relevant glass. Diderot expressed his interest, unlike Voltaire and Hogarth. Its significance was much more theoretical than practical.

Another very concrete consequence of the color theory of Newton and Young was the color print published in London in 1725 by Jacob Christoph Le Blon. In his *Coloritto* he reduced the seven colors of Newton's prism to three fundamental colors, yellow, red, and blue, combinations of which give all possible colors. He selected transparent inks that could mix easily and superimposed prints from three engraved plates, one for each color. The results included famous prints of the head of a young woman and a portrait of Cardinal de Fleury. Le Blon died in 1741, and his student Jacques Fabien Gautier d'Agoty (1717–1785) exploited the process, stole its secrets, and attacked Newton. He did, however, produce an exceptional set of anatomical plates in color.

Color vision has been adopted as a model of a conscious process that involves qualitative, subjective mental states called qualia and can be understood simply on the basis of physiological neuronal processes. Neural coding of color implicates visual pathways from retina to primary and secondary visual cortex with a relay in the LGN. Genetic coding is very important at the level of rod and cone photoreceptors. As we have seen, rods contain a single pigment, rhodopsin, whereas three types of cone have different spectral sensitivities, determined by three different pigments. Their structural genes have been cloned, sequenced, and identified in humans; they are derived from a single ancestral gene by genetic duplication. Here we see a first simple molecular basis for color perception.

The Importance of Color Context

Edwin Land (1909–1991) profoundly influenced research on color perception by demonstrating the importance of the context of a color. In a series of now famous experiments he placed a subject before an experimental "Mondrian" composed of patterns of rectangles of various colors and sizes so that each rectangle was bordered by at least two different colors. He projected light of three different wavelengths on the Mondrian. When he illuminated a single rectangle, for instance a green one, with the three wavelengths together and the rest of the picture in the dark, the rectangle looked grayish white. If he then used a single projector with, for instance, a long-wavelength light, the rectangle looked red. When he lit the whole picture by the three projectors together it looked colored and the formerly white rectangle became green. So perceived color depends not only on light reflected from a surface, but also on adjacent surfaces. The perception of a color depends largely on its context. Changing the relative intensities of the three projectors did not dramatically change the perceived color: it was always seen as green in the case described. There was color constancy. Helmholtz had already recognized that colors appear globally similar whether in natural or artificial light. Land proposed a biophysical retinex theory to take account of color constancy. The brain is free of absolute energies or, in other terms, of defined wavelengths of light. It reconstructs the color of an object in the outside world by extracting a physical invariant parameter from it. For Land this was reflectance. The color of an object depends on the variation of reflectance and wavelength, that is to say, on the relative diffusion (and absorption) of different wavelengths. By examining colored surfaces with three different independent channels, the brain is able to extract from the spectral reflectance of the surfaces in question the invariant color perceived by the subject.

Empathy and Artistic Creation

In her *Homo aestheticus* of 1992, Ellen Dissanayake debated the relationship between esthetic emotion and empathy. For her, creation and esthetic contemplation constituted primarily an empathic relationship. The word *empathy* appeared in 1909, translated from *Einfühlung,* itself derived by Theodor Lipps in 1897 from the Greek *empatheia* and also used by Sigmund Freud (1856–1939). It describes the capacity to identify with another and feel what the other feels. It is to be distinguished from sympathy, participation in another's suffering or compassion, and from Auguste Comte's (1798–1857) altruism (see chapter 2). This last Comte described as an innate human feature consisting of a benevolence toward other members of the community that coexists with selfishness, wherein personal interest is subordinated to that of one's fellows, without religious motivation. Empathy does not necessarily lead to sym pathy, and intentional violence can occur, as it does, for instance, during war.

I shall define art as symbolic intersubjective communication with multiple, variable emotional contents in which empathy appears as an essential feature of intersubjective dialogue. There is empathy between figures within the artwork, but also between the spectator and the figures and thus between artist and spectator. Lipps stated that the vigorous and salient curves of a Doric column brought him joy by recalling those qualities in himself and the pleasure he felt seeing them in others. He claimed that esthetic empathy could be explained by an inner imitation in his consciousness of the very object he observed, his esthetic imitation. Thus, appreciation of a work of art would depend on the capacity of the spectator to project his personality on the contemplated object. Thenceforth esthetic theories included symbolic content and the mimesis of ideas, especially for primitive art. Even abstract art can be included in this approach because its organization and regularity control and replace the chaos of the world. The neural bases of empathy and even sympathy are legion. Apart from neurons in the temporal cortex responding to the expression of emotion, the limbic system of the brain makes an important contribution to the neurology of emotion, either positive toward others (desire, motivation) or negative (anger). Different groups of neurons and the chemical substances they use to communicate with each other, the neurotransmitters (such as dopamine versus acetylcholine), are implicated individually or in combination. Cerebral imaging emphasizes the close relationships between the limbic system (for example, the cingulate cortex and amygdala) and the prefrontal cortex.

We know well that our relations with faces are essential in human social life. We identify familiar or famous people; we judge age, sex, ethnicity, and emotions. We use the face to help understand speech by watching the lips. We make eye contact, try to predict intentions, and so on. There is a well-defined neural basis for these tactics. For example, bilateral occipitotemporal cortical lesions cause loss of face recognition (prosopagnosia). In 1990 Vicki Bruce and Andrew Young proposed a minimal, though very complex, neuronal model for face recognition. It included encoding of structural models, face recognition units, person identity nodes, and recall of names. The recognition of familiar versus unfamiliar faces is automatic and can trigger an unconscious electrical response in the skin in certain prosopagnosic patients. On the other hand, for the identification of a person and the recall of his name, a conscious level, sometimes with a great effort of attention, is required. In 1972 Charles Gross and his colleagues identified for the first time cells in the temporal cortex of a monkey responding selectively to faces, both monkeys' and humans'. Their specificity was remarkable, for the cells did not respond when the eyes of the image were deleted or the image was cut into smaller pieces, nor did they respond to hands or other objects. They even responded differently to images in profile rather than full face, as well as to various facial expressions, such as a yawn, a threatening look, a smile, and the direction of gaze. These neurons were not unlike those described by David Perrett and colleagues in the 1980s and the elaborate cells of Keiji Tanaka (1992). Mark Johnson and John Morton in 1991 studied human babies and showed that a few minutes after birth neonates have an innate ability to recognize features of a schematic face: CONSPEC. But this did not work with a baby of two to five months, by which time a new process is established, that of learning to recognize real facial features: CONLERN.

Many artists have exploited these cerebral mechanisms in representing faces and hands, or the two combined. A striking example is Philippe de Champaigne's seventeenth-century painting *The Repentant Magdalen,* in which he combined admirably her facial expression and the placing of her hands. Both artists and scientists have been concerned with the expression of emotions by the face, or its neural basis. Charles Le Brun (1619–1690) altered Vesalius's anatomical description of the brain to account for Descartes's model, in which the pineal gland "is the place where the soul receives images of passions." The eyebrow, at the level of the pineal, is the part of the face where passions are best visible. "A movement of the eyebrow toward the brain expresses the gentlest of passions; if it turns toward the heart the wildest and cruelest are represented." Le Brun must take credit for this first attempt at neuroesthetics, even if he has since been forgotten by art historians. Continuing the trend, in 1806 Charles Bell described the facial muscles involved in the expression of emotion, and when Guillaume-Benjamin Duchenne (1806–1875) applied faradic electrical stimulation to these muscles and their nerves, he produced the appearance of emotions. Around the same time Franz Joseph Gall proposed his theory of phrenology, in which swellings on the skull were supposed to reflect inner psychological faculties. Contemporary artists such as David d'Angers, Jean-Pierre Dantan, and Gustave Courbet derived inspiration from these experiments.

The capacity of attribution is particularly well developed in humans, as we shall discuss later: it enables us to imagine mental states in others, to attribute to them knowledge, beliefs, and emotions, and to recognize differences and similarities between their psychologies and our own. The mirror neurons discovered by Giacomo Rizzolatti, Giuseppe de Pellegrino, and their colleagues in 1996 in the premotor area of the frontal lobe of the monkey could represent a system for the implementation of attribution. They were active both when a complex motor act (such as reaching for food) was performed by the monkey itself, but also when it was perceived in others (monkeys or humans). From fMRI studies we find that homologues of mirror neurons may be present in or near the so-called Broca's speech area in the human cerebral cortex, which we shall discuss later. They could be involved in imitation, but also in language and—why not?—in esthetic activity. We must realize that a month after birth a baby interacts with its mother through facial expressions of imitation and innovation. A chimpanzee responds to a mirror. In schizophrenics the capacity for attribution is profoundly diminished, as cerebral imagery shows reduced activation of the prefrontal cortex.

Sympathy and Contestation

An artist produces a picture in his imagination, through which he reaches out to other persons. He uses the portrait, and often the self-portrait, in a way that goes beyond a simple self-description, as typified by Rembrandt or Nicolas Poussin.

Research on the neural basis of sympathy using cerebral imaging has recently revealed the perception of pain in a volunteer subjected to painful stimulation and in a closely related partner subjected to the same stimulation. One can distinguish a personal network of pain in oneself and shared networks of pain between oneself and others. A neurobiology of empathy is thus possible, according to Chris Frith. A similar situation exists with regard to violence: an innate mechanism triggers signs of suffering—distress, tears, and cries—which are designed to stop an aggressor's violence and encourage compassion. A sociopathic patient with a violent and remorseless antisocial personality, such as the serial criminal, seems to present with a decreased sensitivity to this inhibitor of violence: there is a prefrontal deficit but no loss of the capacity for attribution. In 2001 Jonathan Cohen's team obtained cerebral images that differed according to whether a subject judged a situation morally acceptable or not. The paradigm was an out-of-control trolley that would kill five people unless a switch was thrown, which would result in the death of only one person (case 1). In an alternative case the five could be saved only if another person was pushed onto the rails and thus killed (case 2). Situation 1 seems morally more acceptable even if the toll is the same. Comparing cerebral images in the two cases demonstrated a difference in the frontomedial and posterior cingulate cortex. So there is a neural basis for moral judgment and, more particularly, sympathy.

The artist exploits these mechanisms. He invites the spectator to share his concept of the world and incites him to contest an intolerable reality. This is Poussin's reference in the *Judgment of Solomon* to the antique precepts of Stoicism as well as to those of the Old and New Testaments. It is the same with political struggles in Théodore Géricault's *Raft of the Medusa,* John Heartfield's *As in the Middle Ages,* Otto Dix's *Dead Soldier,* or Pablo Picasso's *Weeping Woman.* As Claude Lévi-Strauss summarized, "Always halfway between the schema and the anecdote, the painter's genius consists in uniting inside and outside knowledge."

Mental Synthesis and Art's Power to Awaken

We might now attempt a more global analysis of the creation and contemplation of an artwork. This implies active inspection and exploration, as well as progression from one sense to another, from perception to inner vision, hallucination to dream: conscious synthesis. A Darwinian process of trial and error intervenes in contemplation and exploration. The onlooker is not passive before a picture. On the contrary, he explores the work actively, passing from a global view to detailed scrutiny, adopting a projective style. Maurice Merleau-Ponty observed, "Vision depends on looking; we only see what we look at." Examining a painting involves eye movements and gaze. A classic example is the observation of an 1888 painting by Ilya Repin, *They Did Not Wait,* reported by Alexander Luria. He found that eye movements differ according to whether one is looking freely or is under instructions to find something, such as the material situation of the family or the age of the persons. Eye movement is never haphazard, except in the case of frontal lobe damage, such as occurs after surgery for a tumor. So there is, as one might expect, a major contribution of the frontal lobe in the observation of a picture. According to Richard Gregory in 1968: "To read reality from images is to solve a problem." An observer scrutinizes a picture and selects perceptual responses which correspond to his inner expectations. In observing a picture, there is both bottom-up and top-down exploration. An artwork can be conceived of as a coherent, subjective model of reality at the interface of inner vision and outside perception, something like a shared dream. An artwork's power of awaking calls on, among other things, selective autobiographic memories and symbolic or sociocultural representations in their historical context.

Hallucinations invade consciousness involuntarily. It is premature to say whether such processes intervene in the creation or contemplation of art. For example, in his *Disasters of War* of 1820 Francisco Goya illustrated the encounter of reality with nightmare and hallucination. He was not alone. Cerebral imaging of schizophrenic patients during hallucinations shows activation of the limbic system. Jack Cowan recently proposed that the geometric forms of hallucinations are constant, their four principal categories reflecting, after transformation, cerebral retinocortical architecture. It is interesting to note that a rich arsenal of hallucinogenic substances, such as LSD and marijuana, is common among artists, from the Huichol to Henri Michaux.

We shall later develop our model of the conscious global neuronal workspace, which enables us to account for the conscious multisensorial synthesis that takes place when we contemplate an artwork. "To see is already a creative operation that needs effort," said Henri Matisse (1869–1954). The hypothesis that I developed in 1987 and now propose here is that an unexpected, singular synthesis of reason and emotion takes place at a conscious level. Darwinian selection, obviously epigenetic, of global synthetic representations intervenes at this level, enabling the observer to discover the painter's intentions. Seen like this, art becomes a model of social communication that creates an unexpected tension between the constraints of reality and the utopian desires of human society. Art encourages artist and spectator to share a plausible and reconciliatory dream.

Artistic Creation and Mental Darwinism

The creation of a picture is not a simple symmetrical operation, as is its active contemplation. Certainly, the creator possesses the same faculties of attention and selective memorization as the observer, but he also manifests the rarer one of producing public representations. By movements of his hand he projects on the two dimensions of the canvas images of the world he inhabits. This action is by no means instantaneous. In 1960 Ernst Gombrich paraphrased John Constable (1776–1837) when he said that painting a picture was like a scientific experiment. A painting emerges over time from the dialogue of the painter with his canvas, via a complex evolution, or rather by the interweaving of several levels of evolution. Very schematically, we can distinguish at least three evolutions, each of which can be interpreted in Darwinian terms, but with its own specificities. One concerns the elaboration of an intent to portray something or, according to Gombrich, a mental schema. Then there is its progressive realization through the mastery of technique, and finally the ultimate execution of an organized, coherent picture, ready to stand the test of logic.

Edgar Allan Poe (1809–1849) described "a peep behind the scenes, at the elaborate and vacillating crudities of thought—at the true purposes seized only at the last moment—at the innumerable glimpses of idea that arrived not at the maturity of full view—at the fully-matured fancies discarded in despair as unmanageable—at the cautious selections and rejections—at the painful erasures and interpolations," which made the first stages of the creative process an obviously Darwinian mental experience. During this particularly acute, hopeful wait, the artist evokes, dissociates, and recombines images and representations, sometimes without being aware of it, until the ideal form for the original idea emerges.

The role of chance and accident in the genesis of a picture within the mind has often been evoked by painters. Leonardo da Vinci spoke of the power of confused forms, like clouds or muddy water, in stimulating the spirit of invention. Nevertheless, a painting does not come entirely out of the blue. Creativity must work within an existing structure. An artist calls on mnemonic images and representations and an encyclopedia of forms and figures; these have become stabilized in the circuits of his brain in the course of a long process of epigenesis by selection of synapses, which marks each individual in a particular way, like a maternal language. We shall frequently return to this concept of epigenesis (the progressive and coordinated development of an organism in relation to its environment, rather than its being "preformed" and strictly genetically determined). In the sixteenth and seventeenth centuries a central place in a painting was often reserved for the figure of a man, represented naturalistically. Then came elements borrowed from other painted works, especially the artist's own, which helped him discover schematic structures suitable to be adapted to his subject. The individual painting could be integrated in a higher order of things. An artist's creative activity recalls the "tinkering" of the first signs of mythical thought as described by Lévi-Strauss.

One positive aspect of applying the Darwinian model to the genesis of a picture is to define criteria that determine the final decision of the painter. One often invokes reason—the "strategic reason" of Gérard Granger, which relates to the plausibility of goals and objectives. Part of this is the suitability of a commissioned picture to a given theme, the affective reactions that it might potentially unleash, first on the painter and then on the onlooker, and also the painter's own vision of his art. Ultimately a logical coherence between the elements that compose it may emerge in the painter's head, like a revelation, just as the discovery of the solution to a problem "illuminates" a mathematician's mind, according to Jacques Hadamard (1865–1963). So variation and selection of intention intervene at the highest level of the brain, that of reason. It therefore once again seems legitimate to suppose a major role for the frontal lobe in the process of creation. We know that the brain exhibits a large degree of spontaneous activity, the form of which can be regulated selectively by an inner focusing of attention, such as that favored by Michael Posner or Alan Baddeley. We can conceive that in the frontal lobe transitory groups of active neurons form prerepresentations that are maintained in consciousness for a short time, forming a first outline, a mental simulation of a painting.

For Giorgio Vasari (1511–1574) a drawing was the tangible expression and explicit formulation of a notion within the mind, elaborated into an idea. It was the projection of the painter's original outline. The drawings of even the greatest artists, with their hesitations and corrections, trials and errors, show that a new form of Darwinian evolution is in progress between paper and brain. A sketch by the skilled hand of the artist becomes a perceived image, face to face with the picture he had imagined. After this first draft, the painter's hand translates new ideas graphically and incorporates them in the sketch, to complete and enrich it. Then the artist undertakes more experiments, which allow him to explore techniques, invent effective strategies, define mathematical rules, or simply apply methods learned from the old masters, witnesses of the science of the times. The dialogue is continuous from the first rough sketch to the completed drawing, which, in spite of its small proportions, is a good model of the final picture, containing as it does the principal protagonists, their expressions, their relative positions, and subtle indications of light and shade. Finally the painter applies the colors to the canvas. In 1869 Charles Baudelaire wrote that "a harmoniously realized picture consists of a series of superimposed pictures, each new layer giving the dream more reality." Discrete but significant variations reveal the painter's experiments in following the evolution of the painting to its conclusion. There is restructuring through concentration, insistence on the essential, rearrangement of the facts. At each stage the creator becomes a critical spectator, attentive to the resonance of every touch of paint. The sketch is modified by careful attention to form, color, and what Gombrich called pictorial and graphic illusions. There is harmony with the original intention, with repeated trials of coherent logic, rational integration, and reasoned adjustment of the eye. The traces of this unique evolutionary process, the corrections and the superimpositions that distinguish an original from a copy, form a record of the painter's own techniques and his habits in capturing form, applying color, suggesting space. These traces illustrate the characteristics of form and figure that reveal his subjectivity and define his style.

The neural basis of the genesis of such pictures remains enigmatic, but those that control the movement of the hand are better known through the recent work of Apostolos Georgopoulos and Marc Jeannerod. The finely coordinated movements of the fingers that guide pencil lines or brushstrokes are commanded by cells of specialized sensorimotor areas of the cerebral cortex that send their orders, through a relay in the spinal cord, to the muscles. These same cortical areas control direction and orientation of the hand. When a painter steps back from his canvas, his head and eyes change position, but for him the painting, like the rest of his environment, remains stable. Other cortical regions, such as the parietal areas, participate in this invariable reconstruction of the outside world by controlling visual orientation. Lesions cause disorientation, so that a patient cannot touch a target precisely, and his drawing ability declines: he can no longer reconcile his body image with his visual space. Other parts of the central nervous system also contribute to visual guidance of movement, especially the cerebellum, which governs it like an internal clock. The initial programming of a motor act is, however, triggered beyond the motor cortex, in the frontal lobe, close to where we suppose that the first seeds of creative thought blossom and grow. The painting of a picture proceeds progressively. A masterpiece does not happen all at once!

Music and Painting

Human artistic activity fits into a neurohistorical perspective. An artwork is a special cerebral product participating in intersubjective communication: it evolves without definite progress, but with constant renewal. Louise Bourgeois (1911–2010) wrote, "I am not looking for an image or an idea; I wish to create an emotion, that of desire, of a gift, of destruction." Baudelaire spoke of "giving the dream more reality." For Michel Onfray "the artist has the duty to engage in an exchange, to propose an intersubjectivity, to aim for communication." The "sublime percept is the artwork which rivets us with astonishment and admiration by its esthetic efficacy, brutal, immediate, overwhelming . . . after the emotion reason takes over." Like many philosophers and art historians, Onfray does not define "esthetic efficacy."

That is the central question that we should tackle here. In reply, we might suggest a risky but plausible comparison between music and painting.

CONSENSUS PARTIUM AND PARSIMONY

Two features among others mark the esthetic character of a percept and its "efficacy": harmony, or *consensus partium*, and parsimony.

Consensus partium is the coherence of the parts to the whole. This feature is directly related to the fact that an artwork is a human work, an artifact, and more specifically a composition, a special creation limited in space and time, which form the framework within which it grows. This framework can be the physical limits of a painting within its frame, but also that of a musical form with its various parts. We saw that in *Hippias Major* Plato defined beauty as an appropriate formal relationship between the parts and the whole, where the unity of the whole exceeds the multiplicity of the parts. This relationship does not spring from an artist's head like the revelation of some Platonic ideal but results from a long series of trials and errors between the first representation in his conscious workspace and the final work. This idea of "relationship" to which the artist aspires is of value in painting, but it is easier to tackle theoretically and experimentally in music.

The other feature, less well accepted, is that of parsimony. Karl Popper (1902–1994) wrote that "science does not aim at simplicity; it aims at parsimony." Herbert Simon (1916–2001) pursued this reflection and distinguished several levels of science. Basic science describes the world in terms of facts and generalizations and offers explanations of these phenomena: in other words, knowledge and understanding. Applied science establishes laws permitting inferences and predictions as well as inventing and constructing artifacts to implement desired functions. Science is an art: apart from its first imperative, empirical truth, science has an esthetic imperative, which is used commonly by mathematicians. Beauty is perceived in explaining much from little, finding patterns, especially in simple relations, within apparent complexity and disorder. Parsimony is not to be confused with simplicity, which is the opposite of complexity. On the contrary, parsimony denotes the relationship of the complexity of data with the complexity of the formula describing them. For example, (01)* is a more parsimonious formula than the sequence 01 01 01 01 01 01 . . . But why should we pursue parsimony, which is perhaps at the very origin of mathematics? For Simon one human characteristic was an emotional response to the beauty of parsimony, supposed to have been selected by evolution as useful for the survival of the species by reason of the capacity it offers to detect organized patterns in nature.

SYNESTHESIA: RIMBAUD'S SYNDROME

The connection between painting and music is referred to in ancient Greece with the "chromatic" scale of Archytas (428–347 BCE). Certain Greek theoreticians went so far as to consider color as a quality of sound, what we would call *timbre* today. Aristides (530–468 BCE) had gone further: painting lacked moral power, relating only small fragments of life, whereas music, and accompanying dance, had a direct effect on the body and the soul through rhythm and poetry. Aristotle made a vain attempt to quantify color, but one had to await the Renaissance and Gioseffo Zarlino in 1573 for a table of harmonic proportions in music, and François d'Aguilon, who in 1613 extended this table to relations between colors. In 1702 Newton proposed a quantification of the light spectrum as a color wheel, suggesting that the color sequence could be related to Descartes's quantified musical scale of 1650 and that the harmony of colors was analogous to the concordance of sound. It was really Gustav Theodor Fechner in 1876, however, who introduced a scientific concept of colored audition, or *synesthesia,* although it was known to John Locke in 1690. In synesthesia the stimulation of a sensory modality gives rise to a sensation in another modality. It is colored audition, sometimes referred to as Rimbaud's syndrome, for Rimbaud associated colors with letters in his poem "Vowels" of 1883: A black, E white, I red, U green, O blue. Many creative artists experienced synesthesia, notably Olivier Messiaen (1908–1992), who saw music in color, and Kandinsky, who saw painting as music.

Several scientific studies of synesthesia have been undertaken recently. Simon Baron-Cohen and his colleagues found that, of 212 confirmed cases of synesthesia they described, 210 were women, which suggests a genetic predisposition associated with an autosomal dominant gene related to the X chromosome. The most frequent forms were sensations of color triggered by auditory, tactile, or taste stimuli. Other, rarer forms were due to letters, phonemes, or meaningful words. They were not related to learning difficulties but might have been to drugs such as LSD, mescaline, or marijuana. Synesthetes have a spontaneous proclivity to associate sound and visual features, such as color. But they do not confuse synesthetic colors with those seen in the visual environment. Eraldo Paulesu and his colleagues obtained fMRI images of color-word synesthetes and showed that words activated the secondary but not the primary visual cortex, and, surprisingly, not the color area V4. Higher-level prefrontal cognitive visual areas were also activated. Kolja Schiltz and her colleagues demonstrated that potentials for a response to perception of letters was evoked at 20 to 80 milliseconds after the application of a stimulus, whereas the synesthetic response was recorded after 200 milliseconds. Anina Rich and Jason Mattingley asked synesthetes to perform a Stroop task by saying aloud the color in which a letter was written. First they had to determine which color the subject saw for a given letter. Then they presented the letter either with the congruent color (the same as the synesthetic color) or with an incongruent color. They noted a pronounced interference between the displayed color and the synesthetic color, so that synesthetes performed more slowly than control "normal" subjects.

HEARING MUSIC

Music is an organized sound message. Sounds are more or less complex movements, generally vibrations, of the ambient elastic medium (air, water, solid) that produce quantifiable reactions in a receptive subject. To produce vibrations in air takes less energy than it does in water, so amplification is necessary for the passage from the air of the external ear to the watery medium of the inner ear. This is accomplished by a chain of ossicles in the middle ear between the eardrum and the oval window of the inner ear (a 25- to 30-decibel increase). The cochlea forms a spiral of two and a half turns in humans, compared with three in cats, four and a half in guinea pigs, and only one in birds and fish. It contains a basilar membrane along the scala tympani, which bears receptors, the inner and outer hair cells. There are 3,500 inner and 14,000 outer hair cells with, in humans, 30,000 sensory neurons for the processing of sound, including music. Sounds are transmitted from the middle ear to the fluid of the cochlea, then to the basilar membrane, which mechanically propagates waves from the base of the cochlea (high-frequency sounds) to the apex (low-frequency sounds). The outer hair cells are cochlea amplifiers, and the inner ones are true receptor cells, transducing mechanical energy to electrical energy. Movement of the hairs causes changes in the membrane potential. The hairs are of uneven length: displacement toward the longer ones causes depolarization of the membrane, whereas displacement toward the shorter ones causes hyperpolarization (Figure 6). These changes in potential are due to the opening or closing of non-selective ion channels and to movement of potassium ions. These channels are situated at the apex of the hair and connected by a "spring" to the apex of the neighboring hair so that they open mechanically when the hair moves. The relationship between hair movement and membrane potential, however, is neither linear nor symmetrical. The principal transfer pathway for sound information is by type I cochlear neurons, of which the receptive ramifications, the dendrites, are contacted by internal hair cells. Each of the 30,000 type I cochlear neurons contacts a single hair cell, which means that each hair cell can be in contact with ten to twenty nerve fibers. Coding of sound intensity is achieved by increasing the firing rate of type I neurons, whereas coding of sound frequency is by synchronization of impulses with the received frequency. One can demonstrate a characteristic frequency for each nerve fiber and see that they are distributed tonotopically along the cochlea: high frequencies are at the base and low frequencies at the apex. So there is both temporal coding by phase-locking and a tonotopic code, which illustrates partial neural isomorphism between the physical signal and the neuronal output.

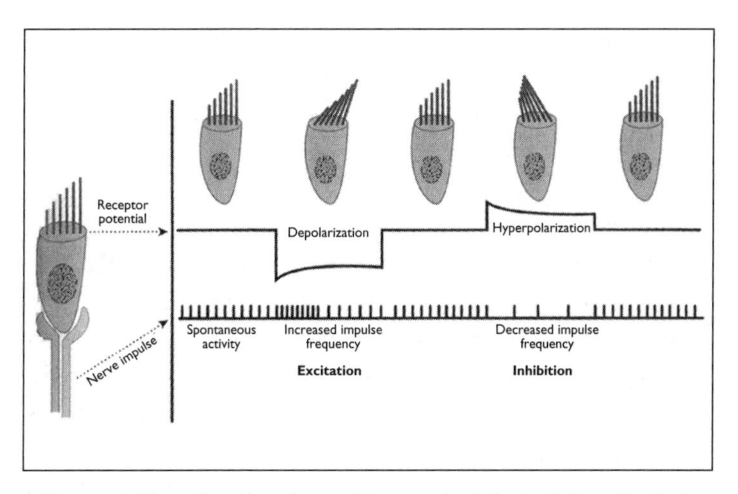

Figure 6: *Transduction of sound into a physiological signal by hair cells of the inner ear.* The cells' spontaneous activity is increased or decreased by the movement of the hairs in opposite directions. From M. J. Zigmond et al., eds., *Fundamental Neuroscience.* Copyright Elsevier 1999; reprinted with permission.

After multiple relays, the information from the cochlea in the auditory nerve arrives at the medial geniculate nucleus of the thalamus, then the auditory cortex. We can recognize, as in the visual system, a ventral pathway specific to the quality of the stimulus (*what*)—for example, a spectral analysis to distinguish vocalization in a monkey or a human—and also a dorsal pathway specific to the localization of the stimulus (*where*). The *what* pathway projects ultimately to the anterior part of the superior temporal gyrus of the auditory cortex, the *where* to the posterior part. Primary and secondary auditory areas exist in our brain with multiple maps of the cochlea, much like those of the retina for the visual system, and they are organized tonotopically (or, rather, cochleotopically). We really do hear music with our brain!

AMUSIA

For a number of years the group of Isabelle Peretz and Robert Zatorre has studied the perception of music as a special cognitive function involving specific neuronal networks. In support of this concept they have identified cerebral deficits that selectively interfere with the recognition of music: amusia, or musical agnosia, distinct from auditory agnosia and language defects, such as aphasia.

Acquired amusia can be produced by cerebrovascular accidents (CVA) and can be very selective. This was the case of Peretz's patient IR, who suffered a bilateral lesion of the auditory cortex and lost her capacity to recognize the national anthem but retained writing skills and was able to recite poems. On the other hand, the composer Vissarion Shebalin (1902–1963) suffered a CVA in his left hemisphere and could neither speak nor understand spoken language but continued to compose, notably his fifth symphony, his masterpiece. The principal deficits concern music itself, the words of a particular melody, sounds in the environment, and voices; they result from lesions of the temporal or frontal lobe. Luigi Vignolo distinguished deficits of melody, due mainly to lesions in the right hemisphere, as opposed to rhythm, mainly in the left. Congenital amusia, diagnosed in early infancy, probably has a genetic origin. The most striking example concerns savant musicians. Their musical ability is exceptional, but they suffer from autism and are socially and mentally handicapped. A case was Blind Tom (1849–1908), a black slave sold with his mother to Colonel James Bethune of Columbus, Georgia. His vocabulary was never more than a hundred words, even as an adult, but at age four he played from memory and without fault a Mozart piano concerto that he had just heard. At six he improvised and at seven gave his first public concert. As an adult his repertoire was of five thousand pieces, all from memory—as he was also, of course, blind. More typical congenital amusia (tone deafness) is expressed by specific deficits, however, such as lack of perception of melody. Peretz found deficits in the recognition of intervals and judgment of consonance or dissonance, whereas timing and judgment of major or minor keys were preserved.

In a similar context certain epileptic fits can be triggered by specific types of music. In 2003 Giuliano Avanzini distinguished epilepsy provoked by classical music, melody, songs, organ music, strings, or jazz, regardless of the patient's musical education. In 1963 Wilder Penfield and Phanor Perot reported that stimulation of the superior temporal gyrus, mainly but not only on the right side, caused specific musical hallucinations that were often so precise that the patient named the piece of music.

CONSONANCE AND DISSONANCE

We have seen that in the plastic arts the notion of harmony, *consensus partium,* is difficult to quantify. But it is easily accessible in musical terms. Physics teaches us that natural musical sounds are often the most complex. Harmonics are pure sounds of frequencies that are whole-number multiples (2x, 3x . . .) of a so-called fundamental frequency. For example if we take the Pythagorean model of three vibrating cords of lengths 1, ½, and 1/3, they produce a fundamental note f and harmonics f2 and f3, respectively. The interval between the frequencies of cords 2 and 3 is the ratio of their fundamentals, 3/2, or a fifth. Further, the first four fifths produce the ancient pentatonic scale of Pythagoras and the Chinese. We can go even further: the thirteenth note of the twelfth fifth is slightly higher than the eighth note of the seventh octave by a small interval called the Pythagorean comma, 5.88 "cents," or a fifth of a half tone. Its equal but artificial distribution over the frequencies of the notes of the twelve fifths defines equal temperament, a concept proposed by Andreas Werckmeister in 1691 and almost universally used in Western music.

Peretz and Zatorre and their colleagues studied patients with cerebral lesions and asked them to judge fragments of Western music for consonance or dissonance, happiness or sadness, pleasantness or unpleasantness. The patient IR, with her bilateral lesion of the auditory cortex, had reduced judgment of consonance and dissonance but could still judge happiness or sadness. Together with Anne Blood and her colleagues in 1999, Peretz and Zatorre used PET scans and showed that when they changed the level of dissonance, there were changes in activation of two main brain regions: the secondary auditory cortex of the superior temporal gyrus, for perceiving dissonance, and the paralimbic (hippocampal, cingulate, orbitofrontal, and frontopolar) cortex, for associated emotional aspects.

Yonatan Fishman and his colleagues in 2001 made detailed electrophysiological studies of responses to various dissonant stimuli, both in monkeys and in humans (Figure 7). The stimuli were chords synthesized from two pure notes at different Pythagorean intervals above C. Ranked from most consonant to most dissonant they were the octave, fifth, fourth, minor seventh, augmented fourth, major seventh, major second, and minor second. Chords composed of tones related to each other by simple (small-integer) ratios, such as the octave

(2:1) and fifth (3:2), were typically judged to be consonant. Those related by complex (large-integer) ratios, such as minor second (256:243) and major seventh (243:128), were dissonant. Dissonant sounds produce beats (frequencies of less than 20 hertz per second) or roughness (20 to 250 hertz). Recordings in the primary auditory cortex of monkeys showed that all the chords produced a high-amplitude, short-latency positive response 28 milliseconds after the stimulus (the P28 wave), mainly in layers III and IV of the cortex (Figure 7) and therefore probably due to synaptic potentials from thalamocortical fibers. Only dissonant chords—but all dissonant chords—produced beats in phase with those predicted by Helmholtz: for example, 13.6 hertz for the minor second and 32 hertz for the major second. It is important to note that the responses were only in the thalamocortical receptive layers III and IV, which form part of the conscious neuronal workspace, to be discussed later. Similar results were obtained in epileptic human subjects by intracranial recordings of Heschl's gyrus of the temporal cortex (the human primary auditory cortex). Dissonant chords produced beats in phase with the beats of the stimulus, a finding that could not be repeated in the secondary auditory cortex. These results demonstrated a physiological representation of dissonance in the primary auditory cortex of both monkeys and humans, indicating that distinction between consonance and dissonance was made early, in the primary sensory cortex. Nevertheless, these studies need to be extended to even higher cortical levels.

The tonal and harmonic context is of crucial importance in listening to music. Marta Kutas and Steven Hillyard in the 1980s compared event-related potentials when a subject listened to a spoken phrase in which alternative logical or incongruous final words were suggested, such as "the pizza was too

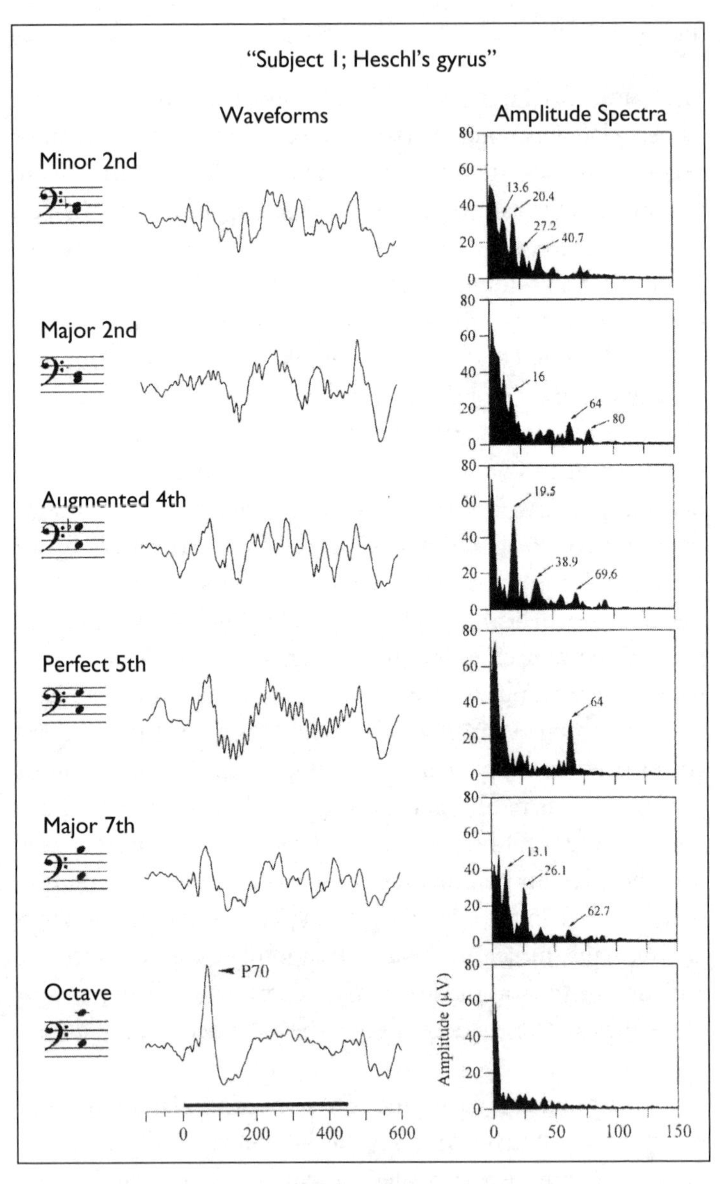

Figure 7: *Evoked responses from human Heschl's gyrus for consonant and dissonant chords.* From Y. I. Fishman et al., "Consonance and Dissonance of Musical Chords: Neural Correlates in Auditory Cortex of Monkeys and Humans," *Journal of Neurophysiology* 86, no. 6 (2001); reprinted with permission.

hot to eat," " . . . to drink," " . . . to cry." The recorded responses varied according to whether the final word was logical or incongruous. The variation was maximal for a negative wave evoked 400 milliseconds after a stimulus, the N400 wave. The more anomalous the word, the bigger the wave. Its amplitude in some way measured semantic expectancy. Mireille Besson extended these experiments to music and discovered that a positive (P600) wave signaled the expected or incongruous ending to a familiar air, such as the "Toreador Song" from *Carmen.* The amplitude of the P600 wave increased from expected to incongruous and from diatonic to nondiatonic. She and her colleagues later distinguished a modification of the P300 wave as a function of the harmonic context and a later (300- to 800-ms) modification related to sensory consonance. In all cases we are dealing with top-down effects on the processing of musical information. These latencies would be those expected for access to the conscious neuronal workspace.

In this context, it is obvious that if music is present in all human cultures as a mode of communication using "patterns of sound varying in pitch and time produced for emotional, social, cultural, and cognitive purposes," as suggested by Patricia Gray and her colleagues in 2001, it may well be present in nonhuman species. Music could have appeared independently in birds, whales, and certain monkeys. Nevertheless, capacity for music differs considerably from monkey to man, particularly in terms of transposition. The macaque can recognize a melody transposed by one or two octaves, but not a half or one and a half octaves, and only in a diatonic scale, not a chromatic. The macaque may be able to distinguish consonance and dissonance, but the tamarin cannot.

I am well aware that our discussion of the physical basis of consonance and dissonance, and its probable representation in the brain ("where" it is), leaves us with a feeling of frustration, for it does not tell us how or why consonance may be interpreted as pleasant and dissonance as unpleasant.

MUSICAL CHILLS: THE EMOTIONAL RESPONSE TO MUSIC

Certain pieces of music provoke in certain persons physiological reactions often referred to as chills, which consist of changes in heart rate, respiration, and the electrical responses of muscles (the electromyogram). In 2001 Blood and Zatorre used PET scans to study subjects who experienced chills while listening, for example, to Rachmaninov's Third Piano Concerto or other emotionally charged pieces. They showed an increased activity in the ventral striatum, the insula, the thalamus, and the anterior cingulate cortex that was proportional to the intensity of their chills, and a diminution in the amygdala, hippocampus, and ventromedial prefrontal cortex: results similar to those experienced when taking cocaine. These are circuits related to the brain's reward system that involve dopamine and opiates. Indeed, the chills response is reduced by naloxone, which blocks opiate receptors. So there is a powerful effect of music on the reward system, which conforms with the model of shared rewards that we shall discuss later, and with our hypothesis of the importance of art, and especially music, in interpersonal communication and the reinforcement of social bonds.

Physiology of Collection and Collectors

Memories of acquired experience remain in our brain throughout life in the form of stable neuronal tracks. They are also transmitted from person to person, from brain to brain, epigenetically. Gestures, attitudes, and language are part of these memories. But memories can also persist and even evolve outside our brain in the form of artifacts, more stable than our perishable cerebral tissue. In 2000 Ignace Meyerson noted that one of humans' characteristic traits is that of producing objects different from what they found in the outside world and which he classified as "works." Works are witness to the most exemplary and stable acquired forms of behavior.

Collecting art is concerned with works in Meyerson's sense, but also at another cultural level, for it indicates that humans are not satisfied with creating works but want to preserve them. He wrote, "Man has also put a value on certain preserved works, he has socialized them." So one might ask if collecting is not at the very roots of a rather surprising domain, peculiar to humans, that of the sacred. Could it be the foundation of the social group that, as a material witness to the creative activity of our brain, confers in return—by what Ian Hacking in 1995 called the looping effect—a powerful symbolic force over our cerebral activity? Collecting would share imaginary significance with the social group and would thus contribute to interpersonal consolidation of social ties. It would immortalize something sacred for future generations.

A collection would be even more: an exceptional record of progress in the evolution of our society—the progress of reason, of course. In his *Pascalian Meditations,* originally of 1997, Pierre Bourdieu wrote: "The world is comprehensible, immediately endowed with meaning because the body . . . thanks to its senses and its brain, has the capacity to be present to what is outside itself, in the world, and to be impressed and durably modified by it." The genetic evolution of species, phylogenesis, has led, through multiple interactions with the environment, including variation and selection, to the formation of our cerebral architecture. This gives us a first understanding of the world and constitutes a first innate representation of it. This genetic evolution is followed by epigenetic evolution of mentality and culture. Works preserved in collections are themselves acquired representations of the world. Their confrontation in a collection contributes to the progress of knowledge, which becomes, as Bourdieu wrote, "a primordial consensus on the meaning of the world." In other words, knowledge becomes an objective. Henri Grégoire (1750–1831) had in a way foreseen the collection as a source of scientific knowledge and its diffusion with the creation of the French National Conservatory of Arts and Crafts. Can one say that the collection incorporates the concept of *habitus* and confers the power to generate and unify, construct and classify? If so, the collection intervenes in two aspects of culture: the sacred and the scientific.

EXPLORE, COLLECT, UNDERSTAND

The behavior of the human collector depends on cerebral functions that we see in other animals, which have an innate tendency to explore the world and an incessant quest for novelty that we can summarize under the generic term *curiosity.* In *The Primordial Emotions: The Dawning of Consciousness* (2005) Derek Denton suggested that, very early in the evolution of species, curiosity appears to have been associated with primordial emotions. In most animal species hunger, thirst, and sex are imperious necessities: life on earth is ensured only because creatures feed and reproduce. For this, neurobiological mechanisms have grown up that incite them to interact quickly with the world. Let us consider thirst. This primordial emotion incites the organism to drink and, having found a water source to quench its thirst, to swallow enough to return its blood to a physiological equilibrium. A few minutes later it is no longer thirsty, for a remarkable phenomenon has occurred. The animal has lost its sensation of thirst before the blood has reached equilibrium. There must be a neuronal system in its brain that produces an inner consciousness of thirst. The same is true for humans. The perception of this need, and its satisfaction, is in a way symbolic because the brain anticipates the physiology. This example may seem far removed from that of the collector. In my opinion, collecting relates to a series of analogous patterns, even if the object is different, intellectually more elevated, symbolically more elaborate, even esthetically more refined.

In the beginning we find primordial emotion: an imperious motivation, a thirst to acquire an object that triggers search behavior, investigation of the world, exploration. Then comes the possible localization of the desired object in a mental map of one's environment, perhaps in an auction room. There follows the creation of a plan of action to reach it quickly while avoiding potential predators, such as competitors with greater buying power. Finally comes the encounter with the object and its acquisition. The collector, even if targeting cultural objects, employs strategies similar to those more "trivial" ones used to satisfy basic human needs.

Exploratory behavior has been the object of numerous experimental studies in animals. It is unambiguously different from more automatic behavior, like simple navigation. During exploration an animal deploys wide-ranging investigative activity such as trial and error, either openly, by interaction with the environment, or tacitly, virtually, by internalization of mental objects. It constantly confronts what it perceives, or imagines, with the real environment or with its memorized inner models of what it is seeking. Its brain acts as a novelty detector, real or imaginary. In so doing, it displays what we might call behavioral flexibility, or curiosity. For this it employs a neural mechanism that we have mentioned several times: the conscious neuronal workspace, which allows access to very diverse representations and enormous behavioral flexibility. Special neurons with long axons, which enable the neurons to act over long distances away from the cell body of origin, allow temporary connections between widely separated areas of the cerebral cortex. Although it may seem something of a caricature, we might look on groups of such neurons almost as curiosity systems. In support of this model, we know that long-axon neurons are a feature of the prefrontal cortex that developed almost explosively with the emergence of humans. In lower species curiosity helps satisfy immediate needs of feeding and reproduction, but in man it also serves more abstract processes and encourages exploration of new domains beyond the physical or biological world, including cultural artifacts, notably artworks. Nevertheless, although curiosity is necessary to begin a collection, it is not enough.

From the outset the first vertebrates possessed the means to evaluate the consequences of their actions on the environment by means of a small but vital number of so-called reward neurons, essential for survival. They are in a way the genetic memory of successful positive or negative experiences from the phylogenetic history of their ancestors, experiences that relate sweet taste to nutritive value or burns to danger. These evaluation neurons enable an organism to maximize its experiences, whether for reward or punishment. There are a few thousand of them situated in the midbrain, and they send their axons widely throughout the brain, notably to the cerebral cortex. They release neurotransmitters, chemical messengers, between neurons such as dopamine, serotonin, and acetylcholine. In humans their evolution has followed several directions. First, they react to novel objects, such as cultural artifacts, which considerably broadens their scope of action within the life of an organism. Second, they enable reflection on the self, auto-evaluation, which provides the organism with the possibility of new inner experiences that are based on the recall of former ones. Ultimately, these reward systems are capable of providing the organism with a means to learn, to anticipate a positive or negative reward, and to plan behavior in consequence.

So if a collector is constantly seeking new acquisitions, that seeking has a neural basis. Reward neurons are involved in this repetitive behavior, which can get out of control and become an addiction. Some collectors lose all sense of the elementary rules of society. Philipp von Stosch was an eighteenth-century German intellectual and antiquarian who became a spy for the British government in order to finance his passion for gems; he had no remorse in stealing from collections he was invited to appraise as a renowned expert. In 1648 Queen Christina of Sweden did not hesitate to confiscate improperly the extraordinary collection of the Holy Roman Emperor Rudolf II a few days before the signing of the Peace of Westphalia. Addictive behavior results from dysfunction of reward neurons in such a way that all activity is turned toward obtaining the addictive substance, whether a chemical or an object of desire, however symbolic. In 1994 Werner Muensterberger wrote that a collector attributes power and value to objects simply because their presence and possession seem to have a pleasant effect on his mental state and provide a way of containing emotions stemming from old uncertainties and trauma, somewhat as religion does. In other words, the collector constantly seeks reassurance through objects. Collection of artworks acquires what Bourdieu called symbolic power. As I have said, it plunges into the physiology of the sacred.

Meyerson identified another activity as unique to humans: not only do they create and preserve, but they classify. Collecting is close to the sources of scientific knowledge, to scientific doctrine that breaks away from nature to build a new reality. Gerald Edelman in 1989 placed this capacity in the center of consciousness. Humans construct a mental scenario in the present in which sensory signals from a nonclassified world are subjected to perceptual categorization. This activity of classification depends on the functioning of a conscious cerebral comparator. The concept of a neuronal workspace includes neurons for evaluation and self-evaluation that would fit this role.

The earliest fossil evidence for classification was reported by André Leroi-Gourhan in 1964; it was based on an excavation in the reindeer cave at Arcy-sur-Cure in France inhabited by Neanderthal man some 35,000 years ago. He discovered a collection of two iron pyrite blocks, a fossil gastropod, and a polypary from the secondary era. We do not know what made someone start this collection. In ancient Greece we find collections of plants and animals and anatomy, as well as astronomical observatories developed around sanctuaries, especially those relating to the Muses. The best known is the Alexandria Museum. King Ptolemy Soter, around 300 BCE, moved Aristotle's Lyceum to Alexandria, providing it with a library of the books of all the peoples of the earth. Likewise, gifts and booty from looting were amassed by authorities. The collections of kings and princes, prelates, doctors, and jurists became signs of distinction and power. In the Renaissance Humanists passionately sought evidence of Greco-Roman antiquity. In the middle of the sixteenth century the phenomenon of the cabinet of curiosities spread through aristocratic society. These cabinets were more like the collections of primitive man than the Alexandria Museum. One found randomly selected natural curiosities, exotic rarities, fossils, coral, petrified wood, animal monsters, gold, ethnographic objects, but also antique statues and modern pictures. Sometimes collections were implicated in scientific controversy. Francesco Calzolari, an apothecary from Verona, assembled a considerable natural history collection with the aim of revising the scientific heritage of antiquity and improving its pharmacopoeia. His illustrated catalogue of 1622 proposed a systematic order instead of a simple alphabetical list. In his catalogue of 1655 Ole Worm, a physician from Copenhagen, criticized several popular beliefs, such as those regarding unicorn horns, which he rightly identified as narwhal tusks. He was a remarkable philologist, and he founded Scandinavian archeology when he published his collection of runic inscriptions. Seen initially as sacred, these collections were progressively secularized. This was a precursor of the encyclopedic spirit and, in a word, of science in the true sense of the term.

The perfection of classification progressed with that of plants by Carl Linnaeus (1707–1778), which was based on an artificial systemization of form, proportion, and situation and introduced binomial nomenclature. The Comte de Buffon (1707–1788) was vigorously opposed, preferring natural organization with increasing complexity, according to Aristotle's "great chain of being." Bernard de Montfaucon, abbot of Saint-Germain-des-Prés in Paris, extended systematic classification to Greco-Roman and medieval antiquities in 1719, and thence to fine art. With Jonathan Richardson's *Discourses* of 1719 and Pierre Crozat's 1741 collection of drawings of old masters, study of artworks became systematic to the extent that Richardson suggested that for the visual arts "connoisseurship" should be considered an authentic science.

Inspired by Francis Bacon, Jean le Rond d'Alembert (1717–1783) and Diderot proposed in the *Preliminary Discourse* to their *Encyclopedia* in the second half of the eighteenth century a generalized genealogy, or tree, of knowledge, their *Figurative System of Human Knowledge.* A remarkable fact for the neurobiologist is that they proposed a division of human science into history, philosophy, and poetry that was based on the three faculties of comprehension, as cerebral as is possible: memory, reason, and imagination. A project debated by the French Convention for a national museum to reflect the aims of the *Encyclopedia* and demonstrate the unity of knowledge was delegated to Antoine-Chrysostome Quatremère de Quincy (1755–1849). It would have an educational aim, "The sight of a collection is above all precepts," and it would offer the citizen an "interesting spectacle." This central museum was never built, through lack of funds, but three independent institutions were created: the Natural History Museum, the Louvre Museum, and the Conservatory of Arts and Crafts. Two hundred years have consecrated this irreparable cleavage, which deprived us of an encyclopedic vision of the world and of an essential confrontation. Must we accept this?

COLLECTING AND CONTEMPLATING ART

There is no a priori limit to collecting, from the collector of buttons or stamps to Don Juan, collector of women. Among the motivations there are, as we have seen, the pleasure of collecting and the pleasure of looking at the collection once collected. There is also a sort of duty to save from destruction memories of the past, of history with its rich heritage of learning. The collection of artworks recalls several of these motivations. It is one of the highest forms of culture, as Jacques Thuillier recently emphasized. It provides an intentionally created environment that encourages enjoyment. Of course, it is witness to history, but its lesson is up-to-date in its human content, which we question as we look.

A painting is artwork par excellence. It has history, style, and meaning. In *The Image and the Eye* (1994) Gombrich pointed out that when making a choice, a collector is faced with "a dazzling and bewildering variety of images which rival in range the creations of the living world of nature." Collecting paintings consists not of simply having a large bank account, getting advice from a professional investor, or an irrational impulse. A collector must make an informed choice, based on wide and enlightened visual experience. In 1699 Roger de Piles in his *Lives of the Painters* distinguished "the curious who form an idea of a painter from three or four pictures they have seen" and who display a naive and irrational infatuation, and the "connoisseur, gifted by talent, reflection, and long experience." A collector should be a connoisseur, but that takes time and effort, and one is never sure of succeeding. Understanding science takes effort, and so does understanding a picture. Therein lies a paradox. We are surprised at this similarity between science and an object of pleasure. The reason is that an artwork is much more than a simple object of pleasure. It possesses potentially a multitude of meanings and a power of evocation attainable only through careful attention. The force of this power of evocation varies with what we agree to call quality. To understand a work means discovering its individuality, its wealth of meaning, and its harmony. Likewise, it means placing it in the context of an artist, a school, a country, and a century.

The art collector, especially of paintings, seeks works that he finds beautiful, even exceptional. Analyzing his behavior may help us better understand what I call quality in a picture. As we saw earlier, according to Onfray, the artist engages in an exchange and proposes intersubjectivity. He distinguishes two stages: first an emotional shock, then a reasonable argument. These two stages recall successive steps in scientific discovery, notably that in mathematics as described by Henri Poincaré (1854–1912) and Jacques Hadamard. Poincaré wrote of initial "illumination" and "character of beauty and elegance" accompanied by "esthetic emotion . . . playing the role of a fine sieve." This fleeting, unconscious effect is followed by a conscious effort of verification and rigorous proof. It seems to me that the collector proceeds in a similar manner in his quest. What happens in his head? Is it the imitative character of the picture—the appearance of an object, a face, an attitude: the mimesis of the ancient Greeks—that attracts him to a given work? Perhaps, but that is not enough. The "esthetic efficacy" of a masterpiece should not be confused with a hyperrealistic copy of reality. We have already seen that Plato offered an important clarification that, contrary to what Onfray thought, did not presuppose any essential dualism in cerebral function. He claimed that beauty is expressed by a formal relationship of appropriateness, where the unity of the whole is more important than the multiplicity of its parts. I think this formula is still topical. In his quest a collector is overcome in a "brutal, immediate, overwhelming" way by the composition of a work, the manner in which it is organized, the musicality of its forms, tones, figures. Our brain perceives, recognizes something physical, consonance or dissonance. We are also sensitive to another feature, parsimony, which we discussed earlier. For Herbert Simon an emotional response to the beauty of parsimony is perhaps useful for survival by the capacity it offers to detect organized patterns in nature. The esthetic collector might also benefit from this exceptional natural quality. He will immediately recognize in a work a certain economy of means revealed as a bold line, a convincing brushstroke, a contrasting juxtaposition of colors, all creating sensory consonance and endowing a work of quality with its own unique harmony.

The immediate appreciation of a composition cannot be satisfied by a copy or an imitation. A work must be an original if it is to have true vivacity or vitality. The collector's eye, or rather his brain, relies on the work's being original. If he hesitates over its being an imitation or copy, doubt arises. Does its overall appearance, its global concept, its design, the formulation of its poetry, coincide with what one expects from the master? On closer examination, do the details suggest a skilled hand, or that of a copyist or novice? What about the boldness of stroke, the preciseness of design, the relation of colors? Can one see corrections in strategic places? Have elements of the original sketch been covered as the painting emerged? Is there a hierarchy in the rendering of the picture so that significant parts are emphasized rather than less significant ones? In such an interrogation of a work Gombrich saw the mind of the connoisseur reiterating the imaginative prowess of the artist. Examining the quality of a work is not recreative, but re-creative. We seek the evolution, the trial and error, of the artist's creative process. The collector sets out to meet the artist in his brain.

Sometimes a connoisseur will immediately recognize the style of a particular artist, like recognizing a familiar face. But sometimes he needs to give thought to identifying the artist. So he compares the painting he has before him with a mental database of works by artists he knows. This is an exercise of memory that is based on a search for similarities and differences between what he sees and what he has already seen. Coordinated firing of widespread neuronal assemblies enables images from memory to pass through his conscious workspace, until he identifies homology, that is, congruence between pertinent features of memorized images and the sensation produced by the work being examined. An artist's name emerges and an attribution is proposed for the picture. In many ways this process of attribution resembles that of the naturalists of the seventeenth and eighteenth centuries such as Linnaeus, Buffon, and Jean-Baptiste Lamarck, who established the first systematic catalogues of living organisms. But there is the added difficulty of the great diversity of an artist's creations in the course of his lifetime, and even at any given moment of his life, as well as the possible disappearance of certain artists' whole production.

In reality the task of attribution often takes on a less favorable aspect. First, the fidelity of memorized images is limited. Photography can help supplement defective human memory, but doubt will remain. Connoisseurship can be acquired by visiting museums and exhibitions and reading scholarly literature: the ferule of the master or tutor is often mentioned by collectors. It is difficult to be an autodidact. Knowledge cannot come from books alone: active perceptivity needs subjectivity. It cannot be communicated entirely by the written or spoken word: it needs the experience of viewing and reward. It also needs critical debate, even if views converge. Attribution through style is hardly a single science. Retrospective examination of the catalogues of Rembrandt's paintings or of the Italian primitives in the National Gallery in London illustrates this abundantly. There are objective criteria, such as the documented origin of a work, irrefutable signs of its having belonged to an artist by an authentic signature, the precise chemistry of the paint, a uniqueness of composition, and a rigorous dating, all of which can help decide, or refute, an attribution. If a work exists in several versions, one of which is in a major gallery, another with a dealer, and another in a private collection, it could be that the artist painted several versions, or his school could have contributed. If the formats or quality vary, and one is large and of excellent quality, would it be the original? Or one of the originals?

Why indeed is the name of the artist so important for the relationship of the spectator with the work? How can his name influence our contemplative approach to the work and participate in the process of re-creation? The lack of care in the presentation of written descriptions of works often disturbs visitors to galleries. This may be deliberate. In some exhibitions all descriptions are omitted and the visitor has to read, if he so wishes, in a separate booklet that he has to carry with him, as if reading about a work, or even being made aware of its content, might have interfered with its appreciation. I believe that a written description is necessary: the reason is the projective nature of our brain. Contemplating a painting is not a passive affair, but demands the active participation of the onlooker. He explores the picture visually, follows its dynamic sequentially, and tries to recognize its underlying organization. Such exploration is close to reasoning, but it demands emotional interpretations and delves into the multiplicity of meanings stirred by the artwork from the immense repertoire of long-term memory. This awakening reveals a huge diversity of experience, of which a major part is subjective and individual. The content of a written description in no way imposes a dictatorial interpretation of the picture; on the contrary, it enriches one's repertoire for understanding it at different levels.

It helps place it in the context of the whole of an artist's or a school's work, and it contributes to a better appreciation of its intellectual quality. A painting needs to be understood. The collector who has selected a picture for its intrinsic qualities as a painting knows well that discovery of the artist's name can revolutionize his relation with an artwork. A name can unleash associations, a semantic maze that can multiply the power of a picture to make one dream without reducing in any way its esthetic quality. A precisely written description enriches the understanding of a work and therefore intervenes directly in the pleasure of its contemplation.

Research on the neural basis of contemplation and collection is still in its infancy. The importance of the frontal cortex has often been emphasized, especially with regard to exploratory behavior and evaluation. This importance becomes even greater when one inspects a picture closely, as one does in contemplation and attribution. The frontal cortex is a recent development in our brain. It is hierarchically at a high level and includes a structural basis for the conscious neuronal workspace. Within it there is a notable interrelationship between cognitive systems and the limbic system, in which emotional responses diversify into an enormous palette of sentiments. Integration of the unconscious processing of images, of searching for names, and of evocation of long-term memory takes place before an artwork reaches the conscious workspace.

Charles Le Brun, Founding Father of Neuroesthetics?

Le Brun and his school, but also his rivals Charles Errard and Pierre Mignard, and his collaborators or disciples Noël Coypel, Charles de La Fosse, Bon and Louis de Boulogne, and Jean

Jouvenet, are the best-known actors in the intense pictorial activity of the second part of Louis XIV's reign. Of the enormous number of works executed during this period the wider audience often remembers only the grand decor of Versailles or a number of official portraits of aristocrats, works that may seem grandiloquent, conventional, and, in a word, academic, with all the dusty and solemn context of that word. This opinion needs revising. With hindsight we can indeed now see Le Brun as a founding father of neuroesthetics. The French Academy of Painting and Sculpture was created in 1648 by Le Brun under the direction of Cardinal Mazarin. It was first and foremost a debating chamber, a school in the ancient meaning of the word, but also an exhibition center with the very political role of the incarnation and diffusion of the Sun King's style. It defined the rules, taught them, and made people aware of them. It was also a place for reflection when the theory of art and its underlying philosophy were developed. Le Brun's speech to the Academy on April 17, 1668, on the expression of passion made clear his own position, which de facto became the official position, according to Julien Philipe in 1994. It helps us better understand this style of painting and perhaps even tackle it from a fresh angle. The art of Le Brun and his contemporaries is surprising, for it lies at the antipodes of painting in the open air, which was to be so successful with nineteenth-century Impressionism. In fact, Le Brun's aim was not to paint from nature but (and here it is of direct interest to us) to reconstruct nature rationally using a formal system to integrate figures and their relations. The aim was not to seek what was special at a given instant in an individual life, but to seek universality both of people themselves and of their history and beliefs. Le Brun's themes, and those of his contemporaries, were traditional, drawn from Catholic orthodoxy and from the Old and New Testaments, but also from history and Greco-Roman mythology—in the context of divine Providence determining universal history—converging finally on the coronation of the monarch by divine right. But through these classical themes Le Brun was a universalist in quite a different way, a revolutionary.

The rationalist nature of Le Brun's paintings paralleled the geometric nature of André Le Nôtre's gardens of Versailles. According to Philipe, Le Brun aimed "to draw up a systematic table of passions, the one after the other, and paint them as they would appear if nature had expressed their essence." His project had certain scientific features: the most mechanistic possible search for truth in terms of definition and physiology of all spiritual passions. Although he did not cite them, Le Brun had read Marin Cureau de la Chambre's *Characters of Passions* (1640) and, of course, Descartes's treatise *The Passions of the Soul* (1649) and his *Treatise on Man,* of which the first edition was published a few years before Le Brun's lecture to the Academy. This can be interpreted as a magnificent account of neuronal cybernetics, which preceded by three centuries many aspects of modern neuroscience; notable exceptions include the role of the pineal gland.

Le Brun's lecture began with a few general definitions. He wrote, "Expression is a naive and natural resemblance of what we have to represent." It was "as much in color as in drawing; it must also be in the representation of landscapes and the assembly of figures." He then tackled the essence of the subject: "Expression is also a part that marks the moments of the soul, that makes visible the effects of passion." There followed a long elaboration in which he addressed "students of painting," adopting almost literally Descartes's *Treatise on Man.* He wrote, "First, passion is a movement of the soul . . . all that causes passion in the soul makes the body react. . . . Action is nothing other than the movement of some parts, and change does not happen by change in muscles; muscles have movement only through the ending of nerves that pass through them; nerves act only through spirits contained in the cavities of the brain, and the brain receives spirits only from the blood that passes continuously through the heart, which warms and rarifies it so that it produces a certain subtle air, which goes to the brain and fills it." He continued: "The brain, so filled, sends these spirits to other parts by the nerves, which are like small fibers or tubes that carry these spirits more or less to the muscles, according to their need to act in ways they are called upon to make." He then mentioned "the little gland in the middle of the brain" where "the two images from the two eyes are assembled into one before it reaches the soul."

As to the soul, Le Brun followed the ancient philosophers in giving it very material attributes or appetites in its "sensitive part": concupiscible for simple passions such as love, hate, desire, joy, sadness, or irascible for wilder or composite passions like fear, boldness, hope, despair, anger, and fear. The king's lesson in neuroscience did not stop there. Le Brun next proposed a curious and highly mechanistic theory on the way in which "inner movements" of a "soul joined to all parts of the body . . . exercise their functions on the face." I mentioned earlier his suggestion that the "eyebrow is the part of the face where the passions are best seen," for it was at the level of the pineal gland. On this theoretical basis he described the expression of various bodily passions. In a lecture in 1680 he emphasized the universality of the resemblance "of parts of the human face with those of animals." In 1872 Darwin mentioned in his introduction to *The Expression of the Emotions in Man and Animals:* "The famous 'conferences' of the painter Le Brun, published in 1667," which "contains some good remarks."

The attempt at synthesis of art and science that Le Brun

and his contemporaries proposed, even if rather different from modern artistic and neurobiological research, deserves to be taken seriously. His "effort at truth" regarding the expression of passions, or emotions, as we would call them today, concerns us. The expression of "simple" passions, which he set apart as "chemical" elements, was taken further by the idea that compounds of these elements of passion are like chemical compounds, as Philipe explained in 1994. This pictorial chemistry of emotion led Le Brun to a universe of clear and distinct interlinked ideas.

Did Le Brun deliberately reject, or not sufficiently emphasize, the detail that distinguishes the subtle moment of sentiment, beyond simple emotion, from within the great variety of facial expressions and attitudes? Two centuries later, Duchenne took up this analysis with the aid of photography. His attempt was just as mechanistic, but he had access to new myological data. He used electrical stimulation to identify the facial muscles that contract and relax to mimic characteristics of the principal emotions. Once again Darwin spoke up: "His works have been spoken lightly of, or quite passed over, by some of his countrymen." He continued: "In my opinion, Dr. Duchenne has greatly advanced the subject by his treatment of it. No one has more carefully studied the contraction of each separate muscle, and the consequent furrows produced on the skin."

Le Brun's heritage did not stop there. His interest in the expression of emotions and their physiology was pursued by Géricault, David, Courbet, and many others. His work exemplifies a rational and universalist ambition, with a scientific clarity well beyond strictly official art.

II
The Good
NEUROSCIENCE AND ETHICAL NORMATIVITY

Since David Hume (1711–1776), philosophy, as well as common sense, has differentiated science from morality. Science establishes facts ("what is"), whereas morality decides "what should be," but many admit that we cannot distinguish what should be from what is. I shall consider whether it is plausible to take an opposite, although perhaps rather surprising, approach and ask whether we can favor what should be from our knowledge about what is. In fact, such a question belongs to a long philosophical tradition, including Hume, Henri de Saint-Simon (1760–1825), Auguste Comte, Charles Darwin, and contemporary ethologists. My idea is to break with traditional philosophical reductionism, which consists of introducing categories that are as artificial as impermeable and which surrounds the debate with semantic and conceptual traps. On the contrary, we need to open the debate to discussion and animate it with some recent, especially cognitive neuroscience, as well as something from the human and social sciences. I shall limit myself to an essentially ontological discussion of the origins of ethics and two closely related themes: first, the natural, essentially neural human predisposition to moral judgment; and second, the evolutionary dynamics of moral norms, both social and cultural, that we have produced in the course of our history. The project is immense and can be presented only very superficially, in merely a few lines, but its aim will be fulfilled if it encourages further reading and discourse, which might just advance our understanding of humans and society.

The Complexity of the Brain

The original hypothesis is that the human brain is capable of elaborating moral judgments, and that it does so. Baruch Spinoza wrote in his *Ethics,* published in 1677, "Everyone judges of things according to the state of his brain." Is this position plausible for the modern neurobiologist, or is it just a whim?

We should first consider the extreme complexity of the human brain. It contains about 100 billion neurons, each interconnected by an average of 10,000 synapses. According to Edelman, these interconnections, counting only those with a fixed connectivity, create a number of possible combinations in the cerebral network that is greater than the number of particles in the universe. If we consider that there is functional flexibility in this connectivity, then we go even further, beyond any fixed limit, which allows for evolution of the social and cultural environment within the brain's competence. Additionally, the connectivity of the brain is not distributed randomly: it is organized specifically for any given animal, being in large part dependent on the power of genes, and also has a random reserve sufficient to ensure, within the genetic envelope, epigenetic flexibility and thus an opening to the physical, social, and cultural world.

As a naturalist, I cannot conceive of trying to examine such a high-cognitive function as moral judgment without introducing the notion of levels of cerebral organization or integration. The brain's organization can be studied from molecule to neuron, from neuron to assemblies of neurons, and further to assemblies of assemblies. It is understandable that socially oriented representations like those involving moral prescriptions call on the highest levels of this hierarchy but are nevertheless rooted in the various lower levels. In an evolutionist perspective, the passage from a given level of organization to an immediately higher level can be conceived of in the context of a model with variation and selection, including the nesting of multiple levels of evolution that participate, each with very different kinetics, in the organizational complexity of the human brain. I shall identify at least four of these evolutionary levels, although there are many more: a genetic one of species, and epigenetic ones of individual history, of culture, and of the development of thought.

Mental Objects

Psychologists and neurobiologists since Donald Hebb have defined internal neuronal states in the brain that represent external or internal objects in a causal manner: an intrinsic function of indication. In neuronal terms, a mental object corresponds to a physical state determined by coherent electrical or chemical activity of a widespread, but defined, population or assembly of neurons. The meaning of a mental object depends both on the body image in cerebral maps and on dynamic activity. It thus seems plausible to consider that social or cultural representations destined for communication between persons, and in particular social conventions and moral rules, utilize such mental objects.

The human brain possesses an almost unique predisposition in the animal world. It can recognize intentions, desires, knowledge, beliefs, and emotions in other people. This capacity to attribute mental states, to put oneself in another's place, allows us to recognize possible differences from or similarities to our own mental states and to plan our actions vis-à-vis others in a way that agrees, or disagrees, with the moral norms that we have internalized. This capacity to attribute, or David Premack's theory of mind, which we shall discuss later, develops in children at around four years of age. It seems to be different in autistic children and certain schizophrenics. It appears to involve preferentially the prefrontal cortex and, according to Paul Ricoeur, plays a critical role in the evaluation of "oneself like another." It thus constitutes one of the fundamental predispositions of the human brain for moral judgment.

The Violence Inhibitor and Sympathy

A long tradition in moral philosophy refers explicitly to emotions like pleasure and suffering. Once again we are dealing with mental states, but specifically in a qualitative way. Whether subjective or passive, pleasant or unpleasant, these fundamental emotions are constantly renewed and are communicated socially by facial expression, as demonstrated by Duchenne, Darwin, and later Emmanuel Levinas. In many animals, including humans, we can detect fundamental emotions such as desire, anger, fear, and distress. The basic cerebral circuits may be different for each of these, but they must be in constant reciprocal interaction. In the context of an organism's exchanges with the outside world, emotions and feelings participate in a self-evaluation to ensure adaptation of behavior toward the world and others, with constant reference to memories of the outcome of past experience.

Anger, violence, and aggression are fundamental emotions. In 1966 Konrad Lorenz, and in 1995 James Blair, proposed a cognitive mechanism, the inhibitor of violence, which intervenes even in children to suppress aggression through nonverbal signals, such sad facial expressions, cries, or tears. Psychopathic children have features of an antisocial personality, devoid of feelings of guilt or remorse in spite of having a normal faculty for attribution. They present a selective deficit of the violence inhibitor, which could be the basis for fundamental moral emotions of empathy and sympathy. We shall return to this theme later.

Internalization of Moral Rules and Social Conventions

A specifically human characteristic is the exceptionally prolonged postnatal period of brain development. This makes adult brain organization critically dependent on the social and cultural environment in which the infant develops. Epigenetic traces of learning, by selection of synapses, are deposited in the developing nerve network. They occur when the maternal language is learned and when beliefs and moral rules are acquired. The adult memory has faster and reversible evolutionary dynamics, involving mainly changes in the efficiency of connections rather than their number. This makes possible the evolution of social representations by innovation, selection, transmission, and storage in intracerebral circuits, as is the case with extracerebral media such as books, images, and artworks. Once these memory objects have been internalized, either consciously or unconsciously, they can act as references for moral judgment that can be updated in working memory.

Consciousness implies a function of the brain, a space for virtual simulation of actions where internal evolution of mental objects can take place, with economy of time, experience, and behavior. This conscious workspace serves to evaluate intentions, aims, and programs of action with constant reference to our perception of the outside world, memories of individual history and earlier experiences and their emotional tone, and internalized moral rules and social conventions. Many attempts have been made to produce models of this, as we shall discuss later, and it is satisfying to note that the neurobiologist's definition agrees with that of the moralist. Indeed, according to Ricoeur, consciousness is "a space for deliberation and experiencing thought, where moral judgment practices hypothetically."

Ethical Normativity

ORIGINS OF MORAL NECESSITY

Having arrived at this stage of reasoning, we might agree that it is not enough to know the place of the human brain in terms of moral judgment to understand the origin of moral rules. Indeed, a search for the origins of moral norms encounters several difficulties. The diversity of cultures that have occupied our world, both in time and in space, and still occupy it raises the problem of relativism of morals, like those derived from philosophies or religions. On this last point Elliot Turiel showed in infants belonging to different cultures, for example the Amish and Orthodox Jews, two different conceptual domains, that of social conventions and that of moral obligations. We shall talk about this later in more detail. Three-year-old children judged as acceptable transgressions of conventional religious prescriptions, such as the day of worship and hairstyle or food taboos, by members of other religions. Transgression of essential moral obligations, such as calumny or physical violence causing suffering, however, was unacceptable. These results suggest that there exists in the child's brain, and in our own, a distinct conceptual domain—a corpus of moral sentiments—of spontaneous moral predispositions, which could indicate the source of a common human ethic. On the other hand, social conventions that involve a symbolic element, religious or philosophical, could vary contingently and neutrally from one culture to another.

SUCCESSIVE LEVELS IN THE NORMATIVE PROCESS

To simplify my analysis of the elaboration of an ethical approach, I, like Henri Atlan, distinguish three closely interdependent hierarchical levels: the individual level; that of interpersonal relations in a given cultural community; and the essential one, that of humanity as a whole. First, whether we like it or not, the survival of the individual and the continuity of the species remain priorities from which no human being can escape. We know that our brain contains systems of neurons involved in major vital functions for survival and reproduction. Philosophers and biologists share common ground on this fundamental principle.

Concerning interpersonal relations, the human race is fundamentally a social species. Jean Piaget and Lawrence Kohlberg demonstrated how the infant first begins to attribute his own point of view to others, then progressively decentralizes and becomes capable of considering points of view other than his own and desists from acting against others. The human brain seems to possess a system for intercomprehension, a critical feature for the regulation of social life. Nevertheless, being aware of others' suffering, even if eliciting sympathy, does not necessarily lead to activity to relieve that suffering. Intentional cruelty is possible. It is even frequent and sometimes systematic in political and military circumstances. Violence can become endemic, spreading from person to person, destabilizing the social group and threatening the affective equilibrium of individuals and their very survival. The elaboration of moral norms for collective harmony then becomes indispensable. It is in a way the price we have to pay to reconcile the enormous representational capacity of the human brain, its ability to judge, and the material conditions for the maintenance of social life.

In the course of history humans have invented systems of values that, as Ricoeur reminded us, are not "eternal" but "depend on preferences, evaluations by individual people, and finally the history of morality." From time to time during the history of humanity, norms have crystallized when individual judgments have created pressures to justify changes of principle. Such selection of moral values has provided humans with their particular natural predispositions at any given state of cultural evolution of the social group. Language and writing have contributed to the elaboration of these rules and their transmission. Such normative syntheses, often expressed by major personalities such as prophets or philosophers—and this is the gist of my argument—have naturally exploited inhibition of violence and sympathy at critical moments of a blossoming, and constantly renewed, cultural evolution. The spread of sympathy and the suppression of violence, which are at the evolutionary roots of humankind, thus constitute, in my opinion, the raw material of the fundamental norms of much of human morality. Together with mutual aid they were placed in an evolutionist perspective by Pyotr Kropotkin (1842–1921) at the beginning of the twentieth century. More recently evolutionary theorists such as David Wilson, Elliott Sober, and Camilo Cela-Conde studied these evolutionary origins of morality in the plausible but debatable context of group selection within human populations.

At the level of humanity, we live in a world where the multiplicity and diversity of cultures, of philosophies, and of religious beliefs seem to obscure an optimistic and harmonious vision of mankind at peace. The early and authoritarian epigenetic imprinting of the child's brain by conventions of the cultural community in which he lives makes the sentiment of cultural, or often religious, identity extremely vivid and particularly stable. In the absence of a rational choice, emotions can be very powerful and lead to violent reactions. Different social conventions seem to thwart the expression of moral obligations that should be common throughout humankind. The alternative of a universal morality imposed by a few people, even if they are well-intentioned, on the rest of humanity seems just as unacceptable, in view of the totalitarian threat to individual liberties that it represents—unless, like John Rawls or Jürgen Habermas, we try to think of ethics not in terms of a particular cultural community but as a theory of society concerning mankind in its entirety.

My idea here is that from monkeys to humans, cognitive functions, especially consciousness and artistic activity, are associated with a major development of cerebral organization. This is manifested mainly by expansion of the cerebral cortex, particularly the association cortex of the prefrontal, parieto-temporal, and cingulate areas. These areas, not specialized for sensory reception or motor command, are closely related to the limbic system, which comprises both cortical and subcortical centers especially involved with emotion and survival. They form the basis for the conscious neuronal workspace.

Models of Social Life and Evolution of Moral Theories

MORAL SENTIMENTS FROM ANTIQUITY TO NIETZSCHE

In the history of Western philosophy, the first written models of social life from ancient Greece referred to a fixed, stable world and aimed to ensure functional harmony. So for Aristotle in his *Politics,* the city was a natural reality and man was "by nature a political animal." The union of several villages formed a city with its self-supporting capacity and its function not only of ensuring survival but of offering well-being. He believed that humans were more social than the bee or other gregarious animals, which we now know not to be true. Humans have a feeling for good and evil, for the useful and the harmful, and in addition use language. The city is a whole, like a living organism. Individuals are integral parts of the city, subject to the whole: alone they are not self-sufficient. In the city human nature is deployed and humans develop well-being. Individuals might differ among themselves, but they have equal access to power in an equitable system. The city is held together by friendship, which might be equated to attachment or social interactions, according to John Bowlby (1907–1990). Aristotle's was the first formulation in psychological terms of sociability requiring active reciprocity. He distinguished three bases for friendship: personal pleasure, utility, and goodness in the form of deep mutual communion. A certain ethical optimism is backed by ethical excellence, which results from education and learning. Ensuing virtues are the right measure and prudence, which, in the contingent, unstable, irregular, and perilous sublunar world in which humans find themselves, seem necessary for meaningful action.

Aristotle's posterity in the Western world has been considerable, for he inspired Thomas Aquinas in the thirteenth century and the medieval Christian tradition. Christian selfless love can be likened to Aristotelian friendship, and charity to the highest levels of that friendship. We should not forget, however, that Aristotle, like Plato, accepted slavery.

From 300 BCE the Stoics, such as Zeno and Chrysippus, developed a moral philosophy with the materialistic view that all reality has a corporal basis. God and the human soul, virtue and passion, are corporal, and humanity is at one with the universe. This theory, which was close to materialistic monism, was different in that the bodies to which the Stoics referred were animated by forces and not simply set in motion, as the atomists held. Like Aristotle, they believed that there is a metaphysical finality to nature. Their empiricist theories depended on sympathy between things and beings. The Stoic lived in sympathy with the universe and nature, striving to match his inner harmony with that of the world. He was not just a citizen of Athens but a citizen of the world and militated for universal brotherhood. The moral being lived in accord with nature, which was profitable, advantageous, practical, and necessary, as well as beautiful and good. Passions, as unreasonable movements of the soul, are on the margins of nature, but still under control: if we can recognize them, we can avoid them through reason. In a way, the Stoics defined a process of cognitive self-evaluation. Appropriate behavior is determined by reason, but a universal reason, itself present in nature, which for the neurobiologist means in the brain. Stoicism is supposed to have had an important influence on the development of Christianity that is due to its universalistic doctrine, without distinction of rank or race, its social ideal of collective happiness and pardon of offenses, and a sort of naturalization of God substantiated as Jesus, the ideal man. Furthermore, such ideas returned with force in the Renaissance with Michel de Montaigne and Nicolas Poussin, and at the French Revolution with Maximilien Robespierre.

The Scottish Enlightenment in the eighteenth century brought forth an optimistic Aristotelian concept of human nature. In 1725 Francis Hutcheson developed a doctrine of moral sense according to which man possesses an innate tendency to universal benevolence, a tendency that includes immediate, instinctive, unreasoned responses to others' misfortunes. Humanity is driven by irresistible compassion. In 1751 Hume spoke of moral sentiment that expresses the emotional reactions of those who judge: we recognize a virtuous action that "gilding and staining all natural objects with the colours, borrowed from internal sentiment, raises in a manner a new creation." In other words, in the expression of emotions the brain adopts a projective function, as Alain Berthoz and I have discussed. In his *Theory of Moral Sentiments* (1759) Adam Smith based the social and moral world on sympathy, a more powerful principle than Hutcheson's benevolence because it calls on the imagination, "the faculty that represents to us the sensations of the other person." Thus, Adam Smith anticipated in a way modern cognitive neuroscientific work on emotion, theory of mind, and the violence inhibitor. The normative process would correspond to a progressive establishment of moral rules and would result from repeated experiences of approval and disapproval that were based on sympathy.

More recently, Friedrich Nietzsche also adopted an antirationalist stance. He reduced ethics to a play of multiple affects underlying each thought. He thus foresaw the idea of the unconscious somatic marker of Antonio Damasio in 1996. Affects form a sort of symbolic language, of which fear is the most important component. Fear is the mother of morality in the face of evil, which Nietzsche saw as "chance, the uncertain, the sudden." To acquire culture is to learn to calculate, to think causally, to anticipate. The result is a will to power, which Nietzsche classified as an affect. Thus, morality possesses an evolutionary character. Respect for morality is a heritage of tradition passed from generation to generation. Acquisition of moral rules is a process of training. Nietzsche followed in the footsteps of Aristotle in the importance he gave to affect in moral judgment, except that for him the feeling of fear differs radically in neural, psychological, and, of course, social terms from friendship or sympathy.

THE RATIONALIST MODEL AND THE SOCIAL CONTRACT

The oldest known model of naturalist ethics stems from the pre-Socratic atomists Leucippus and Democritus in the fifth century BCE, and later adopted by Epicurus. The entire world is formed of eternal and indivisible atoms and the void around them. The human soul is a material body composed of "subtle" matter and of "fine, round, and polished" psychic atoms. Humanity is composed of individuals whose actions have the ultimate end of attaining happiness through individual pleasure, a major innate quality, the beginning and end of human life. Nevertheless, not all pleasure is to be sought out: there is a hierarchy of desire. One has to differentiate among natural, necessary desires, such as quenching one's thirst or relieving pain; natural but nonessential desires, such as that for gastronomic delicacies; and unnatural and nonessential desires, such as those for honor, glory, fortune, or amorous conquest, the last ones of which have to be eradicated. Pleasure is characterized by absence of suffering of body or soul. Peace of mind is attained by study of nature and knowledge of what might frighten man and cause his death. Friendship certainly exists and is even the greatest happiness of life, though in private. Within society there is a need for justice, so the atomists created rules and institutions acceptable to all, constituting a culture in the same sense as conceived by modern anthropology. For Epicurus natural right was a convention to prevent mutual harm. Justice does not exist alone: it is a contract between societies, wherever and whenever. There is relativism of laws and modalities of justice, but this does not remove the need for a social contract to regulate and harmonize social life and interactions between individuals pursuing their own happiness. The Epicurean model of ethics is a minimalist model of the regulation of social life.

Thomas Hobbes (1588–1679) followed a similar pattern, but in the troubled context of war that ravaged England in the early seventeenth century. The cleavage between the natural and the normative cultural artifice is even more marked. Indeed, for Hobbes the natural human condition was not so much a quest for individual pleasure but "a perpetual war of every man against his neighbour." "*Homo homini lupus*"—man is a wolf to other men. The instincts of survival, like those of defense, attack, desire for dominance, hate, and envy, constitute a natural right, but the faculty of thought or reason incites humans to overcome this state of war through the elaboration of a natural law consisting of a social nonaggression pact. This involves abandoning the people's natural right in favor of the sovereign power of a prince and the constitution of a civil society. This society is an artificial human creation needing an effective communication system: language. In Hobbes's eyes humans are a more complex animal than the others because of their mastery of language, which enables them to create pacts and engagements: a civil society with ethical rules. Jean-Jacques Rousseau (1712–1778) took up the idea of a social contract, but in a very different context. In their natural state humans are not wolves to each other; on the contrary, they are driven by simple desires that are in harmony with the physical world. Impelled by instinct rather than reason, they are in a state of innocence, neither moral nor immoral, in which are seeds of sociability. The social contract between one person and other people removes the human from this natural state. Through education humans make the child into a citizen who respects the civil law, which is generally desired and therefore sovereign.

Jeremy Bentham (1748–1832) and John Stuart Mill (1806–1873) and the utilitarians exercised a rational Epicureanism that emphasized the welfare of the majority, in the form of pleasure (Bentham's hedonism) and immediate happiness, or the ideal disinterested happiness of G. E. Moore (1873–1958). For Bentham there existed a moral arithmetic, a felicific calculus, by which everyone undertakes to maximize satisfaction and minimize pain. The most happiness for the most people becomes a moral imperative that is based on a rational evaluation at the level of consequences with no a priori criteria. Mill extended this utilitarian approach by an associationist psychology inspired by Hume. The application of the law of association of ideas, by resemblance, contiguity, or contrast to the moral sciences, made him think that consciousness emerges through associative processes, by the combination of utilitarian motivations and moral feelings. This idea can perhaps be defended if one equates consciousness with moral consciousness. Nevertheless, it seems difficult to accept that the conscious workspace of the human brain, of which the structural and functional characteristics are today quite specific, can simply emerge from a process of association.

Immanuel Kant (1724–1804) countered the empiricist ethics of sympathy and utility with a rational ethics based on the consciousness of duty, derived from absolute and unconditional human reason. Moral duty is an end in itself, expressing an autonomous and reasoning will. The moral law is to be free and reasoning. He wrote: "So act that you use humanity, as much in your own person as in the person of every other, always at the same time as an end and never merely as a means. . . . So act, as if the maxim of your action were to become through your will a universal law of nature." In a way he reformulated the Stoics' principle of universal sympathy, but he presented respect for the human being as a requirement of reason. In so doing he prepared the way for a future secular ethics, but he came up against man's sentiment of moral duty and respect. He even had a presentiment of the concept of evolution. In his 1755 *General History of Nature and Theory of the Heavens* he evoked a theory of the origin of the world involving a chaotic universal nebula that rotated and gave birth to galaxies, stars, and planets. Nevertheless, as Kropotkin emphasized, he lacked the imagination and audacity to include life, and humanity, or to recognize, as did Diderot at the time, that the fundamental system for reasoning had developed progressively in the human brain during the evolution of life on earth. Instead, he lost himself in an imaginary world of angels of divine origin.

EPIGENESIS AND SOCIAL PROGRESS

The application of the concept of epigenesis to humans in society, abandoning the model of a fixed world still present in Kant, is of historic importance, as Georges Canguilhem noted in 1962. Let us look at its great moments.

William Harvey (1578–1657) in his *Exercitationes de generatione animalium* (1651) made comparative observations of the hen's egg and the pregnant mammalian uterus and described in great detail the development of the chick. He proposed that all organisms develop from an egg, passing progressively through several morphological stages to adulthood. There is no preformation, but epigenesis, as Aristotle had already suggested, as an embryo enters successively the vegetative, sentient, and rational souls. In his *Essay concerning Humane Understanding* (1700) Locke extended the theory of epigenesis to human development. According to him, the child's mind is a blank slate, a tabula rasa, which stocks sensory impressions associated with linguistic signs and deploys them in the imagination before being able to understand reality in a rational fashion. Bernard de Fontenelle (1657–1757) pursued this reasoning and proposed in his *Conversations on the Plurality of Worlds* (1686) and *On the Origin of Fables* (1724) an organic analogy between the development of a child and that of human society. The passive world of the child was represented by pagans and savages. Barbarians exemplified the imaginative stage of adolescence. Civilized peoples reached maturity by the development of reason. He claimed, "There will be no end to the growth and development of human wisdom." Fontenelle's and Nicolas de Condorcet's (1743–1794) ideology of unlimited progress, with its idea of infinite human perfectibility, inspired the French Revolution and the U.S. Constitution and, later, social philosophers such as Comte, Herbert Spencer, and Karl Marx. As we shall see, however, the standard model of epigenesis does not lead to a scientific theory of evolution.

Giambattista Vico (1668–1744) in his *Principles of New Science concerning the Common Nature of Nations* (1725) proposed what he claimed was a scientific analysis of the history of civilizations. He distinguished successive ages in the development of human society, which he related to the use of language. First came the age of gods, when social life corresponded to mute acts of religion that helped humans resist the terrors of nature. Hieroglyphics constituted a mute language, and rights depended on divine authority. Then came the age of heroes and the use of articulate language and action succeeding speech. This was the age of heroic and military violence, but also of poetry. Heroic rights depended on force, but with respect for the word. Finally came the civilized age of men, corresponding to the mastery of language. Human rights were based on reason. Blind civic equality was complemented by equity; democracy or enlightened monarchy replaced the heroic aristocrats. Vico considered the possibility of failure of the civilized age and a return to barbarity and a repetition of history, as had occurred after the fall of the Roman Empire. He successfully applied the metaphor of epigenesis to the development of human society and enriched it from two new sources, philosophy, which contemplated truth through reason, and philology: these were sciences of fact and language. Nevertheless, Vico remained a prisoner of the strict theory that development corresponds to a potential created by divine Providence, which is different from true evolution. He believed that the laws of Providence that govern the development of human society are universal and eternal in the fixed framework of natural theology.

Comte also used the metaphor of embryonic development, but he did so in the context of a rather singular natural philosophy. Human history fulfills the natural predispositions of man. This is not strictly epigenesis, for the development of humanity as a collective organism adopted successive defined, almost predetermined states. There again, these states corresponded to the mental development of the individual, but they differed markedly from those of Fontenelle or Vico. In the theocratic state the infant human mind seeks after the first origin, or last fate, of phenomena. He found them in the intentions that animate objects or beings (animism or fetishism), in the actions of supernatural beings (polytheism), or in the acts of a creator god (monotheism). He tended toward anthropomorphism, peopling nature with forces or gods whose actions he based on human ones. Man pursued territorial conquest and slavery. In the *metaphysical* state the adolescent human mind replaced the gods of the theocratic stage with abstract principles, Spinoza's nature, Descartes's geometrical god, Gottfried Leibniz's calculating god, Diderot's matter, and so on. The mind progressed, forsook anthropomorphism, and offered more rational explanations, but it continued to seek the origins of the world. Industry developed, and slavery slowly declined. In the positivity state the human mind became adult. It became relative and renounced absolute theological or metaphysical explanations. Its manner of thinking became that of experimental science, which is based on observation and excludes recourse to useless metaphysical principles. Biology became "organic physics," and sociology "social physics." Politics was conceived and directed scientifically: "Spiritual power will be in the hands of scientists and temporal power will belong to heads of industry." Comte brought Adam Smith up to date and founded a humanitarian religion based on sympathy, replacing "the fictive concept of God by the positive one of humanity." He spoke of a "great collective and socio-affective being" that had "love as principle, order as basis, progress as aim," and "primacy of sentiment over intellect." Science defined the basis of a new culture, a positive philosophy, for which Comte coined the word *altruism,* subordinating selfish instincts to sympathy. Morality became the seventh science, science par excellence. Free, universal, mandatory, and nonclerical education, introduced in France by Jules Ferry in 1882, encouraged intelligence and sociability by developing a positive spirit in the future citizens of a democratic, republican society.

Physiological phrenology emerged as a scientific basis for morality when natural history suggested that there were differences only of degree between human and nonhuman animals in terms of brainpower. Franz Joseph Gall and his assistant Johann Spurzheim produced phrenology charts showing the innate nature of human affective and intellectual faculties, and this led to a new possibility, that there could be a more or less complex interplay of several such faculties in arriving at moral judgments. Comte also divided the brain into organs; the posterior three-quarters or so were concerned with affectivity, inclination, and sentiment, whereas the anterior part housed intellectual faculties, such as observation and combination. He suggested that morality recruited all parts of the brain in concert, corresponding "to the heart, the character, and the mind." Justice is not innate but results from "use of faculties, each enlightened by an appropriate intellectual appreciation of social relations": "intelligence in the service of the heart develops sympathy for the progress of mankind." Comte had the double merit of having made plausible a scientific approach to ethical normativity, and making use of contemporary neuroscience, however fragmentary and often inexact it may have been; contrast him to so many people today who deliberately ignore, or reject, modern data.

A GENERALIZED EVOLUTIONARY MODEL

Lamarck formulated his theory of biological evolution, transformism, on the basis of systematic anatomical observations in his lecture on the systematics of invertebrates in 1801, and he further developed it in his *Zoological Philosophy* (1809). Its extension to the evolution of human society and culture was proposed by two major personalities of the nineteenth century, Spencer and Darwin, each in his own way.

Herbert Spencer (1820–1903) was trained as a civil engineer, and he had been a draftsman for pumps, locomotives, and sewing machines. Very early he adopted Lamarck's transformism in the context of reflections about morality, which he saw as determined by the nature of things. In his *Social Statics* (1851) he founded a naturalistic morality in which paradoxically the idea of social Darwinism was formulated before *On the Origin of Species* was published, in 1859. Later, Darwin not only did not accept the theory but opposed it vigorously. Spencer's *First Principles* (1862) was immensely successful, especially in America. Its popularity coincided with the development of individual, liberal, and competitive thinking in industrial societies in which those who adapted best emerged victorious. This practical philosophy of technoscientific progress and unlimited competition for maximum profit is today that of Western society, whose economic power favors globalization. In *First Principles,* Spencer took inspiration from the embryological work on epigenesis of William Harvey, Friedrich Wolff (1733–1794), and Karl Ernst von Baer (1792–1876). He proposed a law of development that involved the passage from the homogeneity of the egg to the heterogeneity of cells, tissues, and organs in the adult. He broadened this law to evolution in general. It differed from Comte's ideas of preexisting structures and was a more radical concept of epigenesis: "It is proved that no germ, animal or vegetal, contains the slightest rudiment, trace, or indication of the future organism." In fact, there is "adjustment of internal relations to external relations" to maintain "the balance of the functions." The functions are in the domains of biology, intellect, and morality. For Spencer heredity was the reproduction of form and anatomophysiological particularities. Among others, Karl Marx and Friedrich Engels, Trofim Lyssenko, and Piaget retained this Lamarckian adaptationist program, which, in Spencer's mind, beat Darwin's spontaneous variation model. It is remarkable that this surprising and contradictory ideological cocktail of social Darwinism and organic Lamarckism should have had, and still has, such popularity.

In the footsteps of Saint-Simon and Comte, Spencer took up and expanded the theory of the analogy of society and biological organization. For him society was not only an organism but "clustered citizens forming an organ which produces some commodity for national use." In this he paraphrased Saint-Simon, for whom society was "a truly organized machine in which all the parts contribute in different ways to the working of the whole . . . a real being whose existence is more or less rigorous and wavering, depending on whether its organs fulfill more or less regularly the functions they are attributed." Spencer developed the idea of division of labor, wherein hierarchical subordination and political and ecclesiastic powers maintain order and favor salutary activity. For Spencer moral laws were bound up with the general laws of biological and cultural evolution. This secularization of the rules of social conduct followed two principles. First, the reciprocal limitation of freedom of an individual faced with the normal activities of others: this was laissez-faire and noninterference. The second was the proportional reward of aptitude, which justifies rewarding "superior" individuals for their superiority or merit. Spencer was deliberately hostile to any strategy of compensation for deficiency in the weak and the less meritorious. The species would be threatened if advantages were accorded to weaker brethren. The superior take advantage of their superiority by continuous selection. In Hobbes's footsteps, Spencer considered egotism as the principle of moral life, and moral law was composed of "hygienic" rules for the preservation or improvement of society. It "is one that tends ever to raise the aggregate happiness of the species by furthering the multiplication of the happier and hindering that of the less happy." Thomas Malthus (1766–1934) argued along similar lines. Nevertheless, Spencer recognized that there exist altruistic attitudes to biological stability, such as raising children, which might soften egotism, which was for him a final stage in social evolution: universal peace would reign, armies would disappear, and military regimes would be replaced by an industrial system of voluntary, cohesive, and peaceful cooperation. He adopted Epicurean and utilitarian concepts and believed that ethical perfection is related to total adaptation of action to achieving the greatest happiness.

Darwin was a little older than Spencer, and he achieved fame only with the publication of *On the Origin of Species* in 1859. His idea of the descent of species from a common ancestor together with spontaneous variation and selection was the first plausible theory of the phylogenetic origin of species. It remains remarkable that the principles of his theory were broadly in agreement with modern data from molecular genetics. This was a revolution, and more than 150 years later its consequences are only partially accepted by modern society, as Ernst Mayr would say. Its consequences for belief and ethics were immense: the natural theological view of a static world created by God was replaced by a notion of an evolving world that was remote from cosmic teleology or finality and rejected absolute anthropocentrism. The essentialism of a divine design gave way to populationist thinking based on a materialistic process of natural selection interacting with undirected variation and opportunistic reproductive success. Twelve years after *On the Origin of Species,* in *The Descent of Man* (1871), Darwin extended his views to human evolutionary origins and the development of language. He distinguished the evolution of mental faculties that, from fish to monkeys and from monkeys to humans, was manifest by an extraordinary development, particularly of language. There was a parallel with the evolution of languages themselves, for the causes that explained the formation of different languages also explained the formation of different species. Languages or dialects intermingled and crossed, spread or formed subgroups. "We see variability in every tongue, and new words are continually cropping up; but as there is a limit to the power of the memory, single words, like whole languages, become extinct. . . . The survival or preservation of certain favoured words in the struggle for existence is natural selection."

Nevertheless, sometimes Darwin was surprisingly non-Darwinian! He continued to adopt a Lamarckian point of view when he wrote that continued use of vocal organs or language might be inherited, so maintaining the confusion between epigenetic cultural evolution and genetic biological evolution. In chapter 4 of *The Descent of Man* he tackled the subject of moral sense, which he considered the most important difference between man and lower animals. Moral sense could find its origins in lower animals, providing there were sympathy, memory, language, and habits. Some animals possess social instincts to find pleasure in others' company, a sort of sympathy that makes them help others. Animals might also be capable of preserving in their brain memory of past actions and their motives for them, so that they feel regret when social instinct gives way to another instinct. Language might express desires and opinions that could benefit society. Sympathy and social instinct are strengthened by habit. Ethical normativity develops from very basic human instincts. "But as love, sympathy, and self-command become strengthened by habit, and as the power of reasoning becomes clearer, so that man can value justly the judgments of his fellows, he will feel himself impelled, apart from any transitory pleasure or pain, to certain lines of conduct. . . . In the words of Kant, I will not in my own person violate the dignity of humanity." In an even broader perspective Darwin proposed, on the subject of the evolutionary origin of moral rules, that primitive man formed tribes and extended his social instincts and sympathy to the tribe. He rejected moral philosophers such as Hobbes and Spencer who based their views on egotism, and those like Mill and the utilitarians and their emphasis on the greatest happiness. He stated that humans are subject to "an impulsive power widely different from a search after pleasure or happiness; and this seems to be the deeply planted social instinct." Instead of seeking general happiness, humans think of the general good, or the prosperity of the community. On the strictly moral plane, Darwin distinguished higher and lower moral rules. "The higher are founded on the social instincts, and relate to the welfare of others. They are supported by approbation of our fellow-men and by reason. . . . The lower rules . . . arise from public opinion, matured by experience and cultivation."

He in some ways anticipated the distinction expressed by Elliot Turiel between moral obligations and social conventions. He asked how, within the tribe, "so many absurd rules of conduct, as well as so many absurd religious beliefs, have originated; . . . how it is that they have become . . . so deeply impressed on the mind of men; but it is worthy of remark that a belief constantly inculcated during the early years of life, whilst the brain is impressible, appears to acquire almost the nature of an instinct." After this clarification of "internalization" of social conventions, Darwin proposed: "As man advances in civilisation, and small tribes are united into larger communities, the simplest reason would tell each individual that he ought to extend his social instincts and sympathies to all the members of the same nation, though personally unknown to him." "But as man gradually advanced in intellectual power, and was enabled to trace the more remote consequences of his actions . . . his sympathies became more tender and widely diffused, extending to men of all races, to the imbecile, maimed, and other useless members of society, and finally to the lower animals, so would the standard of his morality rise higher and higher." This led humanity naturally to the Golden Rule, "Do unto others as you would have them do unto you." Darwin's Golden Rule originated in the moral evolution that followed in the footsteps of, and even mingled with, biological evolution by a Lamarckian adaptive process.

At the level of biological evolution of man's ancestors, the classic sociobiological theories of evolution of course exclude all inheritance of acquired characteristics. They do, however, admit genetic evolution of altruistic traits at the level of the individual; the recent work of Wilson and Sober I mentioned earlier reintroduced the hypothesis of group selection, which would intervene at the highest hierarchical levels in a collection of individuals interested essentially in themselves. Thus, social structures favoring group selection, and hence cooperation, would themselves be selected positively. Ethical normativity, and especially the Golden Rule, would then emerge naturally by epigenetic extension of biological evolution favorable to social bonds.

In a similar way Kropotkin placed himself in the classic tradition of anarchy by admitting the existence of an objective moral law, in some ways immanent in nature or deducible from it. Opposing Spencer's social Darwinism, he insisted in 1902 that mutual aid is the dominant feature in nature. He based his position both on Alfred Espinas's 1877 study of animal society and on his own observations of nature in Siberia. He noted that in very harsh climatic conditions species survive insofar as they form groups of mutual aid: "The more individuals unite the more they help themselves mutually and the greater are the chances of survival and intellectual development." For Kropotkin the instinctive practice of mutual sympathy served as a point of departure for all higher sentiments—justice, equity, equality, and abnegation—and led to moral progress. Humans' feelings of moral obligation are not of divine origin, but come from nature, partly from animal societies and partly through imitation of primitive man. Wild animals never kill members of their own species, he surmised, and even the strongest are obliged to live in groups. Even if we may doubt this rather "Gaia-ecological" argument, Kropotkin did try to edify a scientific ethic with elements acquired from evolutionary theory, rather than proclaiming the bankruptcy of science, which we hear about in these postmodern times. His program retains some legitimacy even today. In his *Ethics,* written in the early twentieth century, he extended his reflections to the historic evolution of humanity. If humans took mutual aid from the clan to the community, from the tribe to the people, then to the nation, and finally to the international union of nations, in popular movements toward progress, they would improve mutual relations within society. The state should be abolished because it creates divisions in society and forms hierarchies, thus blocking the natural tendency to offer mutual aid. This abolishing of the state would give way to a utopian situation in which a new scientific and realist ethics would give men the necessary force to reconcile in real life individual energy and work for the good of all, thereby opening a new era of cooperation and universal fraternity.

Léon Bourgeois (1851–1925) was opposed to Spencer's individualistic laissez-faire and authoritarian Marxist socialist collectivism. He launched the doctrine of solidarism, a sort of republican morality and duty of fraternity that went beyond the respect of the rights of others, in the form of a positive obligation presented not only as compatible with freedom but conditional on it. Such a concept related to social malaise, which was due not only to personal faults but also to wider causes in which the responsibility of the nation was engaged. Social malaise was akin to Louis Pasteur's theory of contagious diseases, which proved the profound interdependence of all living beings. With his microbial doctrine Pasteur showed "how much each of us depends on the intelligence and morality of all the others. . . . It is a duty to destroy these fatal germs to assure our own life and guarantee the life of all others." Bourgeois referred to Edmond Perrier's 1881 *Animal Colonies* in trying to reconcile science and morality with the constitution of an artificial order that had the force of a biosociological law of union for life. Human societies form groups of solidarity, the equilibrium, preservation, and progress of which obey the laws of evolution: "The conditions of existence of the moral being that these members of a group form among themselves are those that rule the life of the biological unit." Bourgeois was different from Kropotkin in distinguishing in human societies "a new element, a special force: thought, conscience, will." Society lacks justice, is even unjust. There is a duty of solidarity and mutual responsibility. Justice comes from a contract of solidarity, instead of from competition and struggle. This contract is private, individual, and free, equitable for both parties, but also collective and mutual. The state intervenes only to sanction these agreements and ensure their respect. There is "association of science and sentiment in the social task." Humans are united against the danger of natural inequality being reinforced by social injustice.

The Natural Foundations of Ethics

Our discussion has been partial: in the sense of incomplete, and in the sense of biased. But in outlining the evolution of models of society and Western ethical theories, from antiquity to today, the heuristic force and usefulness of the biological metaphor have been revealed. Contrary to a widespread point of view in the human sciences and philosophy, the extension of the biological way of thinking and evolutionary theory to the human and social sciences does not mean the production of repressive totalitarian ideology. Darwin's ethic was not that of social Darwinism but rather that of wider sympathy, just as Pasteur's model of microbial epidemics could form the basis of solidarism. We should remember that various religious zealots for whom moral law is of divine origin, as well as philosophers for whom biology merely reflects sociology, are constantly tearing each other apart and violating the most elementary rules of a common, purely natural morality.

Study of the historical evolution of these various models and their extension to social life and ethics reveals a considerable number of consistent features with universal significance. Reference to human nature, and especially the constraints imposed by invariant characteristics of the human genome, involves first and foremost the concept of moralistic feelings. They can be interpreted on the basis of Jaak Panksepp's 1982 analysis of emotions, which distinguished in the rat, in terms of behavior and pharmacology, neural circuits linked to pleasure, distress, violence, and fear. These feelings would include the friendship of Aristotle, the love and charity of religion, the sympathy of the Scottish Enlightenment, Darwin, and Kropotkin, and the altruism of Comte, although considerable nuances need to be introduced in each case. They relate to reinforcement of social bonds, the breaking of which through, for instance, solitude might stimulate the distress circuits described by Panksepp. Epicurus reflected on pleasure and desire, Hobbes on opposite emotions, such as anger and violence, and Nietzsche put fear and anxiety first. Of course, these all reach the conscious workspace within which we find theory of mind and the violence inhibitor. Also part of human nature is the capacity to reason and self-evaluate, to compare one's mental states with those of others, a faculty described by the Stoics, later by Kant and Ricoeur, and today by our ideas on cognitive science.

The capacity to produce common social representations that regulate individual behavior depends on a process of fixation of moral rules, ethical normativity, and a system for abstraction and generalization found in the prefrontal cortex. Their internalization in the course of development in the child and the adult needs epigenesis and memory, so well developed in humans, their postnatal maturation taking relatively longer than any other animal's. Epicurus, Hobbes, and Darwin emphasized the importance of language in ethical debate for the fixation and transmission of norms—for example, in the elaboration of a social pact. The concept of epigenesis stemmed from the need to explain continuous embryological morphogenesis. It fits beautifully with development of connections in neuronal networks, their selective stabilization, and the formation of a selective imprint of the physical, social, and cultural environment. It frees the brain from fixed mechanistic stereotypes and enables culture to emerge and diversify. It permits the development of society in successive phases, as described by Fontenelle, Vico, and Comte, thus allowing limitless progress. This has curiously been confused with evolutionism by several anthropologists, including Lévi-Strauss in *Race and History* (1952). On the contrary, Darwinian evolutionism did not presuppose an adaptationist program or a preformed design. Group selection, distinct from classic theories of individual kin or reciprocal altruistic selection, allowed the emergence of *Homo sapiens* as a rational and social animal offering his sympathy spontaneously to members of his social group, to all races, and to humanity as a whole, in spite of barriers erected by artificial systems of multiple contingent cultural conventions. It is easy to imagine that diverse societies might have in the course of history independently and naturally adopted the Golden Rule, "Do unto others as you would have them do unto you," without necessarily succeeding in applying it. Evolutionist doctrines commonly give free course to random variation, which refers in neurophysiological terms to imagination, creativity, and innovation. In democratic societies the capacity for what I call "ethical innovation" appears during debates and discussions open to a wide audience. Such discussions are helped, nonexclusively, by numerous concepts developed during the history of Western, as well as Eastern, philosophy. So contemporary bioethical philosophers, such as Anne Fagot-Largeault and Lucien Sève, refer simultaneously to Hippocratic welfare, Aristotelean justice, Bentham and Mill's utilitarianism, Kant's respect for the person, and Bourgeois's solidarity, thus defining a position that we might qualify as philosophical eclecticism.

We must not shackle our discussion to a particular philosophy that might seem to dominate. During the evolution of ideas on social life and normativity, various philosophies have perpetrated borrowing and reborrowing, discovery and rediscovery, of ideas, schemas, and "modules of thought." We should not defend any given philosophical or ethical relativism: to refuse to adopt a single philosophical position, but have an eclectic and open ethical viewpoint, does not signify accepting just any philosophy, just any argument, just any model of society. Our discussion has borrowed various arguments and various modules of thought from various philosophies. This could mark some progress in the way to conceive of ethics that excludes monolithic, integrist reference to any immanent truth, and itself bears features of evolutionism. Humankind has progressively elaborated a certain number of ethical concepts that blend together. Each has a grain of truth as long as it enters into a modular, neurocultural perspective related to the human brain. This combination of modules of thought, a sort of brain mechanism already suspected by Comte, is completed in an evolutionist perspective by constant reciprocal interaction with the social and cultural environment. A selective philosophical eclecticism of this type thus enters into the framework of what we might call ethical, naturalistic, open, and tolerant universalism.

Cultural Evolution

The word *culture* is derived from the Latin *cultura*, which designated cultivating the soil and has come to mean growing plants or micro-organisms. It has been adopted since the sixteenth century to mean education of the mind and the development of intellectual faculties. In the eighteenth century a rift appeared. Kant used *Kultur* in the sense of civilization, still based on intellectual characteristics but implying a hierarchy of civilized over uncivilized. On the other hand, twentieth-century anthropologists, such as Bronislaw Malinowski (1884–1942) and Marcel Mauss (1872–1950), used *culture* to signify arrays of acquired behavior in human societies. A second, more major rift occurred between acquired, sociological culture and innate, biological culture. This distinction lends itself to a conflict between this acquired culture and genetically determined, material "neurobiological" culture. In fact, modern neuroscience teaches us that any cultural representation is produced initially in the form of mental representations, of which the basic neural identity is clear, especially when there is interaction with the outside world. So sociological culture is broadly related to that acquired neurobiologically. The term *neurocultural* seems legitimate without presupposing its being innate.

MODELS OF SOCIETY AND ETHICAL THEORIES

No scientific reflection on the origin of moral rules can separate them from theories of consciousness or society. Western tradition, however, is still largely dominated, even implicitly, by Platonic thinking, which separates a celestial from a terrestrial world. We might recall that for Plato unbridled desire was part of human nature. Humans can attain morality only through an idea of the Good, through the intermediary of a world that links the human soul to the essence of the eternal. This essentialist schema, which we also find in the biblical religions, has the major defect, at least at the heuristic level, of blocking any search for the origin of moral rules in the context of biological evolution and the cultural evolution that followed. On the contrary, a "naturalist" paradigm allows this reflection to benefit from knowledge gathered from the sciences of life, man, and society. It finally liberates the search for the origin of ethics from the many ideological constraints that fettered it for centuries. The classical philosophical distinction, notably by Hume, between what is and what ought to be deserves reconsideration. Paradoxically, Hume thought that the genesis of moral obligation, of artificial virtues, was equivalent to the natural history of human society. Today scientific facts and theories of moral norms come together at the level of evolutionary biology, neuroscience, cognitive psychology, cultural evolution, and the history of philosophy. This is so much the case that it seems legitimate to envisage a true "normative science" integrating the different aspects of the problem in an evolutionist and neurocultural context.

The history of moral philosophies can offer something positive. As Ricoeur suggested, "Ontology remains a possibility today as long as the philosophies of the past remain open to reinterpretation and reappropriation." For the cognitive neuroscientist one of the possible reappropriations is to consider these philosophies as representations of humans and society produced in the course of history by the brains of philosophers. To seek out their invariant features, define their complementarity, and identify the limits of their variability transhistorically leads not to an anecdotal phrenology of philosophical theories but to a search, which we might call eclectic, for authentic natural foundations of ethics. The brains of philosophers and theologians and what they produce belong, like those of scientists, to the natural world, to the biological evolution of the ancestors of humans, to individual experience, to the history of human society.

The biological metaphor has a bad press in the human and social sciences, probably because it recalls serious ideological and political excesses carried out in its name. Contrary to widespread prejudice, there is no question of reducing social man to a genetically determined robot, without culture or history, devoid of sympathy and compassion. Rather, the model of the biological organism will help us discover an array of cognitive systems, initially in the framework of a fixed world, that of the philosophers of antiquity. This array contains moral sentiments, rational evaluation of actions, and the elaboration of an elementary social normativity. At a second stage, that of a developing world, the model of biological organism will enable us to associate epigenesis, cultural evolution, and social progress. In the vision of a world subject to generalized evolutionism, the pertinent extension of the naturalist model should at last reveal the premises of a possible scientific theory of moral normativity.

SOCIAL LIFE

Social life is not specific to humans alone. Many animal species form societies of individuals, exchanging specific stimuli and establishing cooperative links. Social life has appeared repetitively and independently in the course of evolution (polyphyletism) according to modalities that differ from one group to another. In general, social differentiation is more pronounced in invertebrates than vertebrates and is particularly demonstrated by polymorphism of castes. Vertebrates show various forms of antisocial behavior, such as aggressiveness (and even family life can be considered antisocial), but they coexist with "altruistic" tendencies. Some social behavior is peculiar to a species and subject to genetic determinism. Edward O. Wilson analyzed this behavior in insects and considered its generalization to other social groups. In 1975 he formulated a theory of sociobiology that related social conduct in an evolving population with the propagation of the genes that determined it. This theory has provoked numerous criticisms, for it led to attributing "intent" to genes, in an almost Aristotelian manner, Richard Dawkins's *Selfish Gene* (1976) being an example of such attribution. Much human social behavior is acquired through epigenesis. It is thus susceptible to evolution without significant change in genetic heritage.

Within a given culture, representations are transmitted from brain to brain. Among these, beliefs play an important role because they spread as "truths" in spite of the fact that they constitute a provocation against rational common sense, according to Dan Sperber—and of course against any scientific argument. Nevertheless, such beliefs survive and even spread in spite of serious conflicts, as if average human beings relied more on the cultural particularities of their beliefs than the universal evidence of science. In this context it seems legitimate to suggest a Darwinian evolution of mental representations within human societies. Transmissible units, whether Dawkins's memes, Julian Huxley's mentifacts, or Robert Boyd and Peter Richerson's culture genes, might be amenable to replication and therefore transmissible from generation to generation, as well as geographically. One positive aspect of this theory is the definition of the conditions for adoption or rejection (that is to say, selection) of public representations by the brains of individuals in a social group. Some types of representation could, for some genetic traits, present a neutral character, as has been suggested by recent theories of the evolution of species by Motoo Kimura, which would account for the epigenetic diversity of many social processes and, in particular, of the relativism of beliefs and ethical rules.

In this context we might reflect on the natural basis of ethics. Nothing prevents our proposing a neural basis compatible with the 1971 *Theory of Justice* by John Rawls. We can even conceive of the development of a natural ethics along the same lines. It would depend on an equilibrium among individual rationalities within populations of brains and would lead to the elaboration of ethical principles a posteriori on the basis of internal coherence and objectivity. Such a concept of ethics would demand a constant rational critique of standards, especially of beliefs and ideology, and their regular revision to legitimize new forms of social behavior.

SOCIAL UNDERSTANDING AND THEORY OF MIND

The Golden Rule of moral conduct is common to most philosophical and religious traditions and can serve as a typical example of intervention and social understanding in human society. David Premack and Guy Woodruff in 1978 asked, "Does the chimpanzee have a theory of mind?" They asked whether the capacity to interpret one's own and others' behavior by inference from others' mental states, such as desires, intentions, beliefs, and knowledge, was a human particularity. According to them, the expression *theory of mind* is justifiable insofar as the mental states of others are not directly observable by the subject and they must be represented in hypothetical or theoretical form to make predictions about the behavior of others. The term *attribution* may be preferable, as we shall discuss later.

John Barresi and Chris Moore in 1996 placed the problem of theory of mind in a hierarchical evolutionist framework of intentional relations directed toward real or imaginary objects capable of intervening in social understanding. They identified four levels of representation of intentional relations that are based on the distinction between representation of perspectives of self and others, whether real or imaginary. Level 1 involves organisms that have a capacity for anticipation and can distinguish self and others, though without understanding possible similarities between the two. They are incapable of imitation or sharing. They are, for example, rats subjected to classic or instrumental Pavlovian conditioning involving anticipation and reinforcement that develop no coordinated social life. Level 2 is that of numerous organisms capable of sharing representations of self or others, but only in the present and without mutual understanding. The sharing of common activities, with mutual information, extends from the formation of shoals of fish and herds of mammals to the collective croaking of frogs and toads, the polyphonic howling of wolves, and systems of communication by cries of vervet monkeys. Indeed, Dorothy Cheney and Robert Seyfarth in 1990 distinguished several alarm cries in vervets that express different types of fear concerning leopards, eagles, or snakes and triggering different synchronized responses: climbing a tree for leopards, looking into the sky for eagles, looking downward for snakes. But these occur without any collective verification or teaching of infants. Vervets communicate in an elaborate fashion within their group through grunts addressed from lower- to higher-ranking animals, but also vice versa and between different groups. They can evoke responses such as a return grunt or an exchange of gaze, as if a conversation were striking up. At this rudimentary level, however, there is no sign of imitation or mirror recognition, empathy or sympathy. There is no ability to compare the self with others and understand such relationships.

Level 3 involves self-recognition. With an animal's ability for intelligent imitation, as opposed to parrotlike repetition, there appear the interpretation of intentional relations of self and others in a common conceptual system and the imagined representation of self or others as intentional agents. For example, Premack and Frans de Waal noted that chimpanzees in captivity are able to understand the aim of another chimpanzee, imitate it, and annoy it. They are also capable, like whales and dolphins, of providing mutual aid if another is wounded. They show concern toward young handicapped fellows and in general demonstrate sympathy. Especially bonobos, the so-called pygmy chimpanzees, are famous for their affective life and their intense sexuality, as well as their aptitude for reconciliation after scenes of violence, usually with recourse to sex. Furthermore, chimpanzees recognize themselves in a mirror or their reflection in water (Figure 8): they can spend hours looking at the reflection of a spot of paint placed on their forehead or ear without their knowledge. They can even recognize themselves in a distorting mirror and in the face of multiple images.

Level 4 is perhaps attained by chimpanzees insofar as they represent in vivid form, or conceptualize, the intentional relations of others and their own toward others. Like humans,

Figure 8: *A young chimpanzee playing with the reflection of its face by touching the water with its hand.* The chimpanzee can recognize its own image, but its capacity for attribution is limited, compared to that of man. Reprinted by permission of the publisher from Frans B. M. de Waal, *Peacemaking Among Primates* (Cambridge, Mass.: Harvard University Press), p. 84 (Copyright © 1989 by Frans B. M. de Waal).

chimpanzees are capable of introspection and attribution of their mental state to their congeners, and indeed to humans. For de Waal they were capable of deception and developed a genuine theory of mind, which Barresi and Moore disputed, considering it a purely human feature.

Social understanding in humans develops in several successive stages. From the age of two months, and perhaps even earlier, a reciprocal communication is established between a mother and her child, and at the end of the year the child shares mutually coordinated glances with those around him. He communicates by gesture and points to objects or situations in a protodeclarative way. He can use visual and auditory information. In addition, he is capable of representing intentional relations between the first and third persons: humans are at level 2 from the start. In the second year the child reaches level 3, and in the fourth year he develops a true theory of mind. During the second year the child begins to seek out hidden objects, uses imitation and pretence, develops language, and refers to memorized representations to interpret and respond to perceptive events. He uses imagination to compare objects memorized in the past with present reality. He recognizes himself in a mirror, like adult chimpanzees. Babies of less than eighteen months perceive suffering in another baby and will cry in sympathy. A little later their behavior changes and they show signs of trying to comfort another baby who is in distress. "Decentration" has occurred, according to Piaget (1962) and Kohlberg. The baby understands that the sentiments of others can differ from his own and that his attitude can modify them. He can imagine others' mental states and act accordingly (level 3). From twenty-four months old, a child is capable of attributing seeing, wanting, and believing to intentional objects. For Heinz Wimmer and Joseph Posner in 1983 the decisive test of this was false belief. A child can imagine another child not possessing the appropriate knowledge for a new situation with which the former is familiar. In his imagination he compares a representations of his own knowledge with that of another (level 4). As James Baldwin recognized in 1894, understanding the self develops in parallel with imagined, but real-time, understanding of others. There is an obvious relationship between self-knowledge and empathy or sympathy.

In 1995 David Premack and Ann Premack used an elegant video animation (Figure 9) to demonstrate that children aged ten months or more attribute "human" intentions and aims to very simple mobile objects such as images of balls that either "caressed" or "hit" each other. The baby coded caresses positively and hits negatively. In another scenario a ball helping another to escape from a frame was evaluated positively, but hindering it from escaping was seen as negative. A baby attributed an internal cause to intentions and understood reciprocity; for instance if A caressed B, then B should act positively toward A. He also appreciated the difference between a ball bouncing well and one bouncing badly. The very young infant thus possessed spontaneously a system of moral values, favoring cooperation and sympathy and even esthetic concerns. Autistic children develop serious cognitive dysfunction affecting social communication, affective contacts, empathy, and sympathy. According to Simon Baron-Cohen, Alan Leslie, and Uta Frith, autistics lack theory of mind. They cannot infer information in the first and third person, and are thus reduced to level 1. Several attempts have been made to identify neural correlates of theory of mind by cerebral imaging (Figure 10). On the basis of psychological tests involving the recognition of terms specifying particular mental states, Baron-Cohen and his colleagues in 1994 reported increased activity

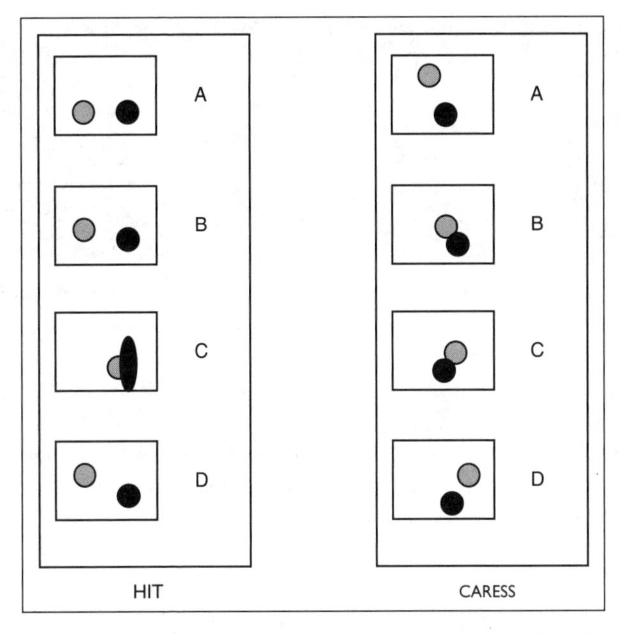

Figure 9: *Computer images used by David Premack and Ann Premack to evaluate empathy and sympathy in children.* From David and Ann Premack, *The Baby, the Ape, and Man* (Paris: Odile Jacob, 2003); reprinted with permission.

in the right orbitoprefrontal cortex. Vinod Goel and his collaborators in 1995 used tests of knowledge about twentieth-century humans and their motivations, and showed selective activation of the left prefrontal cortex. In spite of the differences, both sets of experiments implicate the prefrontal cortex in theory of mind. This is not unexpected, as it can be seen as the most recently evolved part of the human brain.

We saw earlier that in 1966 Konrad Lorenz described an inhibitor of violence in animals, part of the nonverbal cognitive communication mechanism further described in 1970 by

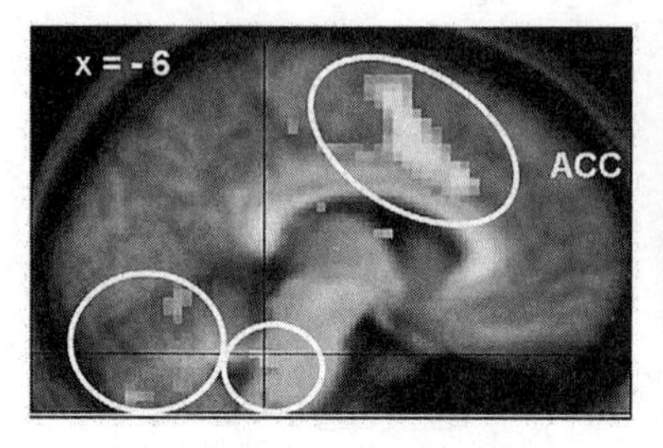

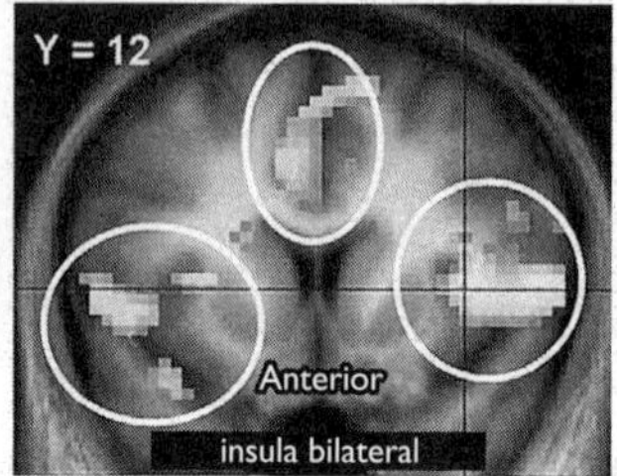

Figure 10: *Functional MRI of evaluation of pain in oneself and in others.* Images were compared when a painful stimulus was applied to a woman or her partner. There is overlap between the two series of images, emphasizing our capacity to share others' emotions. ACC = anterior cingulate cortex. From G. Hein and T. Singer, "I Feel How You Feel but Not Always: The Empathic Brain and Its Modulation," *Current Opinion in Neurobiology* 18 (2008): 153–158; copyright © 2008, reprinted with permission from Elsevier.

Irenäus Eibl-Eibesfeldt. The victim produces signals, for example, of submission (by exposing its throat to a dog), which inhibited the attack behavior of the aggressor. As I mentioned, James Blair extended this concept to the human child in the form of a model for the development of moral awareness. His theories favored a model of spontaneous activation of the violence inhibitor during development. Between four and seven years a child reacts to a sad expression, cries, and tears by ceasing violence. What might be called moral emotions follow: empathy, sympathy, guilt, and remorse. The violent act has been inhibited. Though autism seemed to result from selective alteration of theory of mind, psychopathic children present, according to Blair, a selective deficit in the violence inhibitor. The psychopathic child shows no emotional reaction to distress in others, a fact that conforms with this view. He is violent and aggressive without remorse or guilt, although he knows that he has caused suffering, which shows that his mind is intact. Various authors, such as Hans Eysenck in 1964 and Gordon

Trasler in 1978, supported a theory that a fear of punishment for transgressing moral principles will encourage the child to develop moral behavior.

Elliot Turiel produced a fundamental study on moral and social reasoning in infants and adolescents, demonstrating that they could distinguish those judgments of moral necessity that were compulsory and noncontingent from those that were related to social convention and were voluntary and contingent. He studied children from distinct fundamentalist religious communities, Amish-Mennonites, Dutch Reform Calvinists, and Conservative and Orthodox Jews. They were asked precise questions about rules of conduct concerning day of worship, baptism, concerning covering of women's (or men's) heads, and concerning food. He found that "nonmoral religious rules were judged to be relative to one's religious group and contingent on God's word. Thus, it was thought that religious rules were not applicable to people of other religions, and that members of their own religion would not be obligated to follow the rules if there were nothing in the Bible about them. Judgments about the moral rules entailed a different kind of connection to religion. It was thought that members outside one's religion were also obligated to follow those rules, and evaluations of the moral acts were not judged to be dependent on God's word." These children therefore distinguished between compulsory moral rules and more conventional nonmoral rules.

Coevolution of Genes and Culture, and Cooperative Behavior

As I discussed, social life is not specific to man. It is particularly well developed in insects. In bees alone it has appeared independently a dozen times or so. There could have been evolutionary convergence in certain aspects of social life, but the genetic mechanisms involved could differ from one group to another. Two main mechanisms have been postulated to account for the development of altruistic behavior: individual selection (kin or tit-for-tat) and group selection, spotlighted by Wilson and Sober in 1994.

KIN SELECTION

The idea of the intervention of kinship in the evolution of altruistic behavior stems from an in-depth analysis of the social life of hymenoptera, the ants, wasps, and bees, performed notably by Edward O. Wilson. In particular the very varied bee group, the *Apoidea,* which contains around 20,000 different species, demonstrates several stages of evolution of social life, from solitary species, to species forming colonies of two or three individuals, to the honey bee, which may gather in a swarm of tens of thousands. Bees differentiate progressively from male drones, to sterile workers specialized in collecting pollen and nectar, to the egg-laying queens. Workers and queens emerge from the same eggs, but the larvae are brought up differently, which produces two different epigenetic phenotypes. When the concentration of a chemical signal, or pheromone, secreted by the queen, trans-9-keto-2 decenoic acid, diminishes, the workers construct special cells where the larvae are raised on royal jelly, a secretion particularly designed for the purpose, and produce queens instead of workers. The old queen is driven out of the hive and forms a swarm. One of the most remarkable features of this epigenetic differentiation of workers and queens, in addition to their different fertility, is their entirely different behavior and, logically, their neural organization. The workers contribute to the collective work of the hive, building and cleaning cells, searching for and collecting food, consuming pollen, feeding the larvae, as well as performing various dances. They spend much of their time "doing nothing," however, patrolling or resting and constituting a reserve force ready to intervene either in an overall maintenance process or in time of crisis.

Superimposed on this epigenetic variability is considerable genetic diversity. The males are *haploid,* meaning that they only have one member of each pair of chromosomes, like sperms or eggs, because they hatch from unfertilized eggs, whereas the female workers and queens are normally *diploid.* The altruism of the workers, then, is directed toward their diploid sisters, and the haploid males do not show any particular social behavior. In 1963, on the basis of his observations, William Hamilton proposed a theory of genetic evolution of social behavior. His theory of kin selection proposed that there was close genetic proximity between two sisters, and altruism was more likely when there was close kinship. Kin selection happens at the individual level, thus providing a plausible basis for the evolution of social life in hymenoptera.

TIT-FOR-TAT SELECTION

This theory of the evolution of cooperative behavior was developed in 1984 by Robert Axelrod, on the basis of computer games, to answer the question asked of evolutionists: how could cooperation occur among selfish individuals without a central authority? There were two supplementary questions: what strategy could prosper in a heterogeneous environment containing other individuals using a great variety of more or less complex strategies; and under what conditions could such a strategy, once established in a group, resist the invasion of a less cooperative strategy? The problem could be asked in the form of the "prisoner's dilemma." The game concerns two players who privilege their personal interests in the absence of any obligation to cooperate. Each player has two options, either cooperate or go it alone. Each has to choose without knowing the other's decision. It is better to go it alone rather than cooperate unless (and therein lies the dilemma) both players go it alone, in which case they will do less well than if they have cooperated. When the contest was repeated over a long period of time with many players, more altruistic strategies fared better than greedy ones. Axelrod suggested that this was a possible mechanism for the evolution of altruistic behavior by natural selection. The best deterministic strategy was found to be "tit-for-tat."

The extension of tit-for-tat theory to biological evolution, however, raises a number of extra problems. For the strategy to develop at the genetic level, it is essential that the partners are able to recognize each other, to recall earlier encounters, and to receive a reward for cooperation or punishment for defection. In concrete terms, there should be access to a fixed meeting place, such as a coral reef or other territorial facilities for meetings between neighbors.

GROUP SELECTION

In their famous 1979 article "The Spandrels of San Marco and the Panglossian Paradigm: A Critique of the Adaptationist Programme," Stephen Jay Gould and Richard Lewontin challenged the use of the word *adaptation*, retaining only genetic drift or the constraints of form as explanatory principles for biological evolution. Wilson and Sober were also successful with their "Reintroducing Group Selection to the Human Behavioral Sciences" (1994), opposing the former point of view and rehabilitating adaptation at the level of the social group and not just of the individual. According to them, selfish individuals might win over altruists within the same group, but groups of altruists could win over groups of egoists: groups evolved like adaptive units. Wilson and Sober's proposal was that natural selection operated on a nested hierarchy of units: group selection enveloped individual tit-for-tat or kin selection. The social group, a social organism composed of individuals, would be homologous to a biological organism composed of cells. So we can conceive that a mutation causing hyperaltruistic behavior at the level of the individual, such as increasing the productivity of a worker bee in the hive, could be rejected at the level of the group. A Stakhanovist bee would break the work rhythm and disorganize the activity of the hive. A hive in which the bees work in harmony would beat a hive in which the activity is less well organized. The colony has become the vehicle of selection. Genetic evolution can then occur through selection between colonies or groups.

Wilson and Sober applied the model to two examples: groups in which the dominant form, or *allele,* of a gene coding for altruistic behavior is expressed only in a brother-sister lineage; and groups in which reciprocal tit-for-tat altruism occurs. They analyzed in detail several cases of the second application in the context of the metaphor of all being "in the same boat." In a boat with several oarsmen it is not necessarily the best one who makes the boat go better, even if he exerts himself. It is the one who adapts best to the optimal rhythm of all the whole crew. These simple mathematical models apply to particular cases. What is missing is a general theory.

In support of the still controversial theory of group selection, several examples of specialized cognitive capacity seem to represent adaptations at the group level. Coming back to bees as an example, in 1995 Thomas Seeley described different behavioral strategies to ensure appropriate running of the beehive. Workers choose among several sources of food and fly to that richest in sugar. The greater the richness of the source, the more bees are recruited through a dance that indicates its direction and distance. As the source runs out, the threshold for the dance decreases. Young bees adapt the speed at which they transform nectar as a function of the speed at which older bees collect it. The number of workers recruited for the reception of nectar increases in proportion to its influx to the hive. This recruitment is signaled by a tremble dance inside the hive, the duration of which increases with the richness of the source. In general, a collection of behaviors exists in the hive to maximize the success of nectar collection and its exploitation. When workers behave optimally, their hive stands a higher chance of survival than less efficient hives.

Tribolium castaneum is a small beetle that is easy to rear in the laboratory, so its evolution over several generations is possible to study. In 1976 Michael Wade followed the size of forty-eight populations of sixteen adults over nine successive generations while selecting for large populations, small populations, or none at all. Selection was very effective. Large populations contained an average of about two hundred individuals, and small ones about twenty. This difference was conserved for three years after the end of selection, pointing to a genetic basis. This was not an additive genetic effect, nor was it due to genetic interaction. Two mechanisms were possible. The first was cannibalism, for adults eat eggs, pupae, or larvae, and there are important genetic differences for this behavior. Migration was another, for speed of migration increased with population density. In either case there was group selection leading to apparent differences in fertility in the group, unaccompanied by change in individual fertility.

Egg production by chickens has also been studied for strictly economic reasons. To increase productivity, hens can be raised several in the same cage, but they become aggressive and engage in cannibalism. Selecting cages containing less aggressive hens, with therefore a lower mortality, can increase egg production by 160 percent. Group selection has more than just a theoretical interest.

DECISION MAKING IN AFRICAN BUFFALO

In 1996 Herbert Prins studied the behavior of herds of buffalo, particularly during their moves in search of new pastures within their territory, which could be from five to one hundred square kilometers, and of which they had a mental map. Before setting off, members of the herd evaluated the area to which they would move, using their memory of previous visits and the rewards the area had offered. A collective decision by the whole herd was reached before starting out. This took place at the end of the afternoon. Around five o'clock the herd lay down to rest. Only a few females, the permanent members of the herd, would get up for a minute or so, gaze in a given direction, then lie down again, so that of a herd of 950, only five to fifteen were standing at any given time. Around six o'clock the whole herd would stand up and within minutes the buffalo would all begin to walk in the same direction. This move represented not the orders of the head buffalo, but a vote by the permanent members. Prins concluded, by recording their direction of gaze, that the females who stood up must have been deciding on the direction to take for the next grazing. Thus, there was sharing of information concerning a collective decision affecting the survival of the group.

The studies I have just described highlight global rules within groups requiring coordination but not necessarily individual sacrifice. In 1989 Larry Michaelson, Warren Watson, and Robert Black compared success in resolution of problems among students working alone or in groups. The result was unambiguous: group decisions scored higher than individual performance, even among the best-performing students. John Timmel and David Wilson obtained a similar result in 1997 for a word-search game.

NORMATIVE SOCIAL INFLUENCE AND COLLECTIVE DECISION MAKING

Two examples of anthropological research illustrate the evolutionary significance of normative social influence, particularly at the moral level, in the context of group selection. Wilson and Sober analyzed the social behavior of Hutterites, a fundamental religious sect that appeared in Europe in the sixteenth century and emigrated to America in the nineteenth. They practice community of goods, share their property, and cultivate selflessness. Nepotism and retaliation are condemned as immoral. Personal property is precluded in favor of the community, and not just one's own relations. For Wilson and Sober this social conduct had a considerable advantage in terms of group selection. In 1997 Christopher Boehm undertook a similar analysis of egalitarianism in hunter-gatherer societies. Chiefs had little authority. Adult autonomy was regarded positively, and there was power sharing through a reversal of the dominance pyramid seen typically in primates. There was a common desire to smooth out differences in the phenotype, whether genetic in origin or not, in order to attenuate the effects of individual selection within the group and increase the effect of the group itself. In this way there was intensification of differences between groups. Reinforcement of moral rules of sharing made a group that subscribed to them more competitive than those who did not. The debate on the importance of group evolution in human evolution is open!

III
Truth
A NATURALISTIC CONCEPT OF THE WORLD

How Have We Arrived at a Naturalistic Concept of the World?

Western philosophy grew from that of the ancient Greeks at the end of the seventh century BCE. Among the first were the Milesians: Thales (ca. 624–548 BCE), Anaximander (ca. 611–547 BCE), and Anaximenes (ca. 570–500 BCE). As Geoffrey Lloyd recounted in 1970, this was a time of important technological progress, the end of the Bronze Age and the beginning of the Iron Age. Classification developed from observation and rationalization. These early philosophers discovered nature and distinguished between natural and supernatural, and they then avoided the supernatural. Thales declared that gods were everywhere, but he left them there. He discussed earthquakes, which were due to agitation of the water in which the earth floated. Although referring to a Babylonian myth, he differentiated the universal from the accidental and the contingent. In his rare surviving texts Anaximander wrote that living creatures were born of water as it was evaporated by the sun: humans originally resembled fish. Was that an ancient myth or an evolutionist prediction? Anaximenes described air as the fundamental substance of our soul. In general, the early Greek philosophers and scientists attempted to discover the material basis of what the world is made of, as Aristotle said. They were independent of scripture and elaborated a first naturalistic explanation of the world. Their tools were objectivity, discussion, and comparison of theories so as to find the best. Jean-Pierre Vernant noted how discussion preempted ritual: Greek culture sprang from the city and was opposed just as much to religion as to royalty.

Representation and Knowledge: The Major Stages since Antiquity

At the end of the sixth century BCE Pythagorism contrasted with Milesianism by proposing that numbers are primordial: numbers are the elements of all things, and even the sky is a musical scale of numbers. Numbers are harmonious with the order of the world. This myth still exists among certain modern mathematicians and physicists, and Pythagorism has brought to science a mathematical representation of objects and phenomena.

In the fifth century BCE Heraclitus, Parmenides, and Empedocles asked whether the world is in perpetual change, within certain limits, caused by underlying tensions, as in the case of the bow and the lyre. Heraclitus doubted the witness of the senses, and Parmenides called on judgment by reason. Empedocles went further, in proposing that worldly objects are composed of roots: earth, water, air, and fire, mixed in different proportions, associating and dissociating by attraction and repulsion. There was evolution by selection: monsters that could not resist disappeared and the human race was born. Further, Empedocles proposed an empirical theory of perception and knowledge that the atomists later adopted. Objects release emanations like microscopic particles that enter pores in sense organs. Knowledge results from the attraction of like by like. The atomists—such as Leucippus in the first half of the fifth century BCE and his student Democritus (about 460–370 BCE)—postulated that everything exists in the form of atoms and void. The differences between physical objects are explicable in terms of modification of the form and arrangement of atoms. They are indivisible and solid and capable of unlimited combinations. The atomists extended Empedocles' theories and rejected finality. Their first principle was what was natural: men and animals are born of the earth. The soul is corporal, composed of the same fire as celestial bodies. It is mortal, and its atoms decompose with the death of the body. Sensations are seen as touching at a distance. Sensations, and thought, depend on penetration by simulacra detached from visible objects. Humans can know only what has penetrated their bodies. The atomists were skeptical about "truth." What is important is that Democritus stated that the brain, as guardian of thought and intelligence, contains links to the soul. Here is a genuine naturalist theory of knowledge. The Hippocratic physicians of the fourth and fifth centuries BCE saw disease as a natural phenomenon, far from any magical force and divine or demoniacal intervention. Epilepsy was due to discharge in the brain.

Plato retrogressed somewhat from Democritus. He returned to the Pythagorean world of numbers and saw "ideas" as objects of true knowledge. Explanation of the world is impossible through experimentation, and possible only by contemplation of ideas. All science is mere reminiscence, whereas ideas are innate. He believed in teleology, and finality in nature, ideas later denounced by Spinoza, Diderot, Darwin, and Freud. Plato adopted Empedocles' theory of simple primordial bodies, with simple geometrical forms, however, and so anticipated mathematical theories of physics and chemistry. Furthermore, his preference for reason over simple sensation was a major addition to science.

Aristotle rehabilitated observation and experimentation with his criticism of Platonic ideas, which he qualified as empty words and poetic metaphor. He sought causes, of which he considered four: material, formal, efficient, and final. Matter aspires spontaneously to form, then form to efficient and final causes, which we might see as function today. In his *History of Animals* he established a classification of animals that is still basically valid today: invertebrates without red blood, vertebrates with red blood, oviparous reptiles and birds, and viviparous mammals. He recognized genera and species and proposed a hierarchy of form and function, from simple to complex, a *scala naturae,* with man at its summit. But he ignored evolution. The living world was static. Species and their forms were fixed. Contrary to Plato, who saw the soul as separate from the body, Aristotle considered the soul and the body as inseparable. He distinguished three levels. First, the vegetative soul, common to all living beings and ensuring nutrition and reproduction. Next, the sensitive soul for receiving form, rather than matter, and imagination, the image persisting after the object disappears, in memory or dreams. Finally, the rational soul, possessed only by man, ensuring the formation of concepts and reason. He subdivided the practical intellect, the receptacle of images, and the theoretical intellect at a higher level. Broadly, he recognized sensation, concepts, and reason and introduced some intellectual gymnastics in the distinction between true and false.

A few centuries later, in the fourth century CE, Nemesius made a serious attempt, on the basis of Galen's experiments, to relate the organization of the brain with its capacity for representation. This model divided the soul into motor, sensory, and reasoning functions; imagination, reason, and memory were in the anterior, middle, and posterior ventricles of the brain. We had to await Descartes and his *Treatise on Man* to see a renewal of scientific discussion of the relationships between body and soul. Posterity sees him as a dualist, but Descartes also anticipated modern developments in cognitive science by proposing a connectionist model. He postulated that the "machine" of the brain was composed of microscopic tubes, like "little nerves" through which passed animal spirits. He also suggested a hierarchical organization of this lattice of tiny tubes, from the muscles and sense organs to the "concavities of the brain," then to the pineal gland, then to the cerebral cortex, which he left blank in his sketches, but where he might well have placed the soul. Similarly, he distinguished sensation, a "movement of the brain, as in animals," perception, mixing spirit and body to stimulate the pineal gland, and judgment, reason proper to the soul. In reality Descartes remained ambiguous, even contradictory, throughout his philosophical works on the subject of the precise relationships of body and soul, beyond the decisive role of the pineal gland, which was soon challenged by such eminent anatomists as Nicolas Stenon and Thomas Willis. This could well have been due to fear of the ecclesiastical and political authorities.

A decisive stage in our understanding of the process of

how we represent our environment came with Lamarck's *Zoological Philosophy,* published in 1809, when he was sixty-five. He rejected a "fixed" world created by God, with man at its summit, for a "transformist" concept of the living universe, in which he conceived of a slow progression from one species to another, accompanied by increasing complexity of the nervous system. "Reasoning" man, however, occupied a special place. During the evolution of the brain, its capacity for forming mental objects or representations of sensory input from its environment grew. Lamarck considered thought a physical act that develops progressively with the emergence of inner feeling or, as we might say today, a conscious workspace. The mechanism by which "circumstances influence the form and organization of animals," or inheritance of acquired characteristics, corresponds to an Aristotelian empiricist mechanism of evolution of species:

Matter → Form 1 → Function 1 → Form 2 etc.

Lamarck formulated for the first time evolutionism founded on solid, major observations. For him "nature gives to animal life the power to progressively acquire a more complex organization." Changes in environment resulted in changes in needs, which created changes in habits: organs then changed according to two rules. The law of use and disuse stated that "more frequent and sustained use of a given organ fortifies the organ little by little, and it develops and grows." The law of inheritance of acquired characteristics stated that "what is acquired by the influence of circumstances . . . is conserved by the race by reproduction for new individuals who derive from it."

In *On the Origin of Species,* not published until 1859, Darwin followed Lamarck's lead but went much further. With his theory of natural selection he proposed the first model of a plausible mechanism for biological evolution. He was destined to become a clergyman and was brought up on William Paley's *Natural Theology* (1802) and *Principles of Geology* by the essentialist creationist Charles Lyell. In 1831, aged twenty-two, young Darwin began a journey around the world, during which he collected and classified a large number of species. He was struck by the distribution of the animal populations he encountered in South America and by the geological environment of its past and current inhabitants. In 1837 he wrote up his notes on the transmutation of species and in 1844 wrote a draft of *On the Origin of Species*. The published version contained a suggestion that hereditary variability of domestic species might be "affected by the treatment of the parent prior to the act of conception." Darwin noted that there was an analogous variability in nature so that it became arbitrary to distinguish species or varieties. He emphasized the importance of the struggle for survival and finally introduced the principle of natural selection as a mechanism for "preservation of favourable variations and the rejection of injurious variations." He added the notion of sexual selection, which depended not on a struggle for existence but on a struggle between males for the possession of females. Geographical isolation played a decisive role in the continuous process of natural selection and divergence of characters: *natura non facit saltum*. After a chapter on the difficulties of his theory, such as the rarity of transmitted variations, diversity of instincts, and the great perfection of organs, Darwin concluded by proposing a "grand natural system" with a classification founded on genealogy in which embryology revealed the prototypes of each great class. This was distinct from any "plan of creation," gave new foundation to psychology by the gradual acquiring "of each mental power and capacity," and threw new light "on the origin of man and his history."

Darwin took a major epistemological diversion in proposing the three steps of his selectionist model. First was an exponential growth of populations of living organisms, but with limited natural resources. Then there was spontaneous genetic variation of individuals in populations, and inheritance of these variations. Finally came natural selection through a struggle for existence by individuals whose survival depended on their inherited constitution. Taken together, they resulted in a gradual change in the form of organisms and therefore their capacity for representation.

Originally conceived to account for biological evolution at the genetic level, this model expanded to include the evolution of representations in conscious space. In 1855 Spencer proposed that coordination centers were inserted between sensory and motor "groups," thus contributing to increased connectional complexity and therefore functional integration. For John Hughlings Jackson (1835–1911) there was evolution of highly organized lower centers to less well-organized higher centers throughout life, which was marked by a change from the automatic to the voluntary. J. Z. Young, in *A Model of the Brain* in 1964, relied as much on cybernetics as experimental observations of the octopus when he conceived of the organism as a homeostat that maintains its own organization in spite of changes in environment, choosing a particular action from among a variety of possible ones. An organism does this as an adaptation insofar as it enables self-maintenance and survival. In this sense the organism is, or contains, a representation of its environment. So the world of a fly is different from that of a mouse or a man. With evolution the capacity for representation expanded from the physical and biological environment to the social and cultural. With cultural evolution, externalization of representations developed from internalized representations, through epigenesis and memory.

Evolutionary epistemology extends and generalizes the Darwinian model to other human activities. It is the expression of a philosophical movement that, as Donald Campbell affirmed in 1987, takes into account humans and the human brain as products of biological and social evolution and sees in evolution, even in its biological aspects, a process related to knowledge. The paradigm of natural selection for increased knowledge can be generalized to epistemic activities such as learning, thought, or advances in science. From this point of view, the best scientific laws possess neither analytical nor absolute truth.

Karl Popper was the principal modern author to have developed and argued the variation-selection model in science, beginning in 1934 with *The Logic of Scientific Discovery.* He wrote: "We choose the theory which best holds its own in competition with other theories; the one which, by natural selection, proves itself the fittest to survive. . . . A theory is a tool which we test by applying it, and which we judge as to its fitness by the results of its applications." In *Conjectures and Refutations* (1963) Popper showed that "we actively try to impose regularities on the world. . . . This was a theory of trial and error—of conjectures and refutations." Reasonable belief, as proposed by David Hume, was replaced by reasons for accepting or rejecting a scientific theory. Popper completed the evolutionist perspective by suggesting that it implied multiple mechanisms intervening at several distinct levels, articulating with each other and interlinked hierarchically so that at each level there was a form of selective retention.

In chapter 6, "Of Clouds and Clocks," in his book *Objec-*

tive Knowledge (1972), Popper contrasted clouds, disorderly and unpredictable physical systems, with clocks, "which are regular, orderly and highly predictable in their behaviour." For a Newtonian physicist, if man is a complete or closed physical system, he becomes entirely predetermined: there is a total absence of freedom. For Popper, on the other hand, freedom is "the result of a subtle interplay between something almost random or haphazard, and something like a restrictive or selective control." If so, freedom would correspond to the edge of randomness introduced by variability at each hierarchical level.

He generalized the evolutionist paradigm of trial and error to multiple nested organizational levels in higher organisms. Organisms engage in strategies for resolving problems, a concept extending to development of knowledge by conjecture and refutation. Knowledge that is validated by its agreement with "facts," and is therefore "true," is gradually organized into a tree of knowledge, constantly updated by new facts. Though for Popper consciousness was simply interaction between numerous control systems, he (somewhat curiously) continued to defend a dualist standpoint, distinguishing at all cost physicochemical states from mental states.

For Popper "natural selection of hypotheses" by conjecture and refutation leads to a growing "tree of knowledge." The expanding nature of scientific discovery means that it appears as a rich network with multiple nodes where new rational objects are constructed. As he noted, objectification, the calling to rational order of the world through science that resulted, reduced teleology to causality and in the end contributed to the survival of the species, thus satisfying the Darwinian principle. Popper found himself confronted with the body-soul problem, however. Certainly, he rightly considered consciousness an interactive control system, but he was no longer consistent when he refused to identify mental state with physical state and postulated two sorts of interactive states. This inconsistency dissolved with the appearance within the cognitive sciences of a program that indeed established a reciprocal causal relationship between perceived state of consciousness by a subject and the state of his brain.

The reformulation of these ideas by Campbell in 1987 emphasized the blind, rather than random, character of the fundamental process of production of variations, which are independent of the environmental conditions of their occurrence. The occurrence of individual trials is independent of the solution. Variations that follow unsuccessful trials are not the corrections of these trials. The global process is fundamental for all inductive success, for all increase in knowledge and adjustment of the system to the environment. It comprises, in addition to the mechanism introducing variations, coherent processes of selection and mechanisms to preserve and propagate the selected variations. Campbell distinguished rather arbitrarily ten levels where such processes of variation and retention could occur. Among these figured habits and instincts and the solution of nonmnemonic problems, such as locomotion of the paramecium or William Ross Ashby's Homeostat of 1948, a device capable of adapting itself to its environment. Then he included thought guided by vision or supported by memory, like the unconscious work of a mathematician working on a mathematical problem, mentioned by Poincaré in *Science and Method* (1908). Or that of the artist who for Gombrich (1959) proceeded by sketches and corrections. Or the exploration of social substitutes: there was economy of cognition when the results of exploration by trial and error by a member of a group were learned by imitation by other members of the group. Finally, we could mention language and accumulation of culture.

Many scientists have considered the production and epigenetic transmission of cultural entities in human society, such as the memes of Richard Dawkins in *The Selfish Gene*, or the public representations of Dan Sperber in 1996. Several adopted a Lamarckian model of social transmission that could be legitimately replaced by a model of social communication dependent on a process of variation and selective retention by the brain, particularly the transfer from short-term to long-term memory. This model includes recognition of novelty, the competence of the receiver in relation to the collection of memorized representations and prerepresentations (intentions and anticipations), the pertinence of the sender in relation to the prerepresentations and to the expectation of the receiver, as well as the power to generate (creativity).

Such an enterprise is valid only if, like Popper and Campbell, one distinguishes several levels of selective variation and retention in the neurofunctional organization of the brain. To the molecular level (variability of the genome, evolution of the organizational plan of the nervous system) must be added levels of connectivity of neuronal networks, levels of first-order assemblies (understanding) and higher-order assemblies (reason), all subject to epigenetic variability with shorter and shorter time scales. The definition of elements subject to selection and the understanding of mechanisms of selection engaged at each level represent an extremely difficult multidisciplinary task that is still mostly undone.

The Objective World

According to Joëlle Proust (1995), "A philosophical theory is naturalistic when it recognizes as legitimate only those objective undertakings and explanatory principles normally recognized and adopted in the natural sciences. A naturalistic theory illustrates the genesis of knowledge beginning at its most recent state without being blinded by the inevitably temporary and refutable character of the explanatory hypotheses that it proposes." The problem is to explain the capacity of a neuronal system to represent an external, or internal, state purely causally. For Fred Dretske in 1988 there existed two levels of causal explanation: first, a correlation between an external and an internal state, which was a representation of the external one, then a causal connection between the internal state and a behavioral output or action on the world. So objective facts, independent of the observer, form the material basis for something to signify, or indicate, something about something else. Any naturalistic theory postulates the preexistence of an external reality that can be divided. First, there is a physicochemical or biological world possessing a defined structure, its own organization, not created, or labeled, or intentional. Then there is a human world created by humans, which includes social interaction with other human beings and their mentalities and contains cultural aspects created by humans in the form of artifacts and labeled by them, such as industry, artworks, and writing. The world *not* represented by humans must be distinguished from the world *already* represented by humans.

The theory I propose depends primarily on the predispositions possessed by living organisms to represent the outside world. First, an increasingly complex neuronal architecture, determined by the constraints of a genetic envelope, a product of biological evolution, offers an opening to more and more extensive and abstract representations. This includes the capacity for learning and memory, fundamental emotions, conscious deliberation, and judgment. Second, widespread, but topologically defined, populations of neurons of which the electrical and chemical activity is coordinated and coherent, and which codify, or indicate, a meaning in a defined context, are mobilized in space and time. Third, the abandoning of the input-output scheme held for so long by cybernetics and neurophysiology, in favor of a projective style that Alain Berthoz and I have described. This manifests itself by the exploratory behavior of various animals, by attention, and by the visual gaze that precedes perception and action—in other words, by the spontaneous formation of patterned states of activity, or prerepresentations, phenomena that might enter into hypotheses about the world in the context of mental Darwinism.

In reality the brain's interactions with the outside world are realized along two diametrically opposed lines. One is centrifugal, in the projective sense, with prerepresentations, and is analogous to Darwinian variation. It results from transitory spontaneous activity in various populations of neurons, acting as generators of diversity by forming random combinations of preexisting neural systems, selected both by biological evolution and by epigenesis during development. The other is centripetal, brought into play directly by interaction with the outside world through mechanisms of sensory perception. Correlation of the activity of percepts is determined by the characteristics of objects in the outside world, interaction with which updates memory tracks in the brain. These two directions lead to selection of the appropriate prerepresentation via evaluation systems, in particular reward and emotions. The reasons for this selection include survival of the organism, effective communication, and a harmonious social life. Through it a relationship between an object in the outside world and its neural representation is established. Nested representations can pile up at various levels of organization in the nervous system. A basic classification distinguishes an individual's personal "private" representations from "public" ones transmitted from one individual to another. Of the private representations, some are conscious and others unconscious. Of the public ones, in 2000 Dan Sperber recognized first-order representations, empirical facts stored as "true" in encyclopedic semantic memory, and higher-order representations of representations, such as scientific hypotheses, or more normative representations, such as beliefs, moral rules, or laws.

In 1995 Alain Connes and I proposed that it was legitimate to postulate a hierarchy of evaluation and selection, which in the case of intelligent "thinking" machines included a basic level ("I lose, I win, I lose, I win"), an evaluation level taking into account memory of losing and winning strategies, and a level of creativity that recognized novelty and identified whether a new prerepresentation related to an existing reality. The same word can have different meanings in different contexts (for example, the word *representation*), and the same object or person can be designated by hierarchically different words. Noam Chomsky raised these two objections in "Language and Nature" in 1995. Both difficulties can be resolved by a connectionist approach, the first to its semantic context through lateral interactions, the other to its hierarchical nesting of concepts through vertical connections. This first outline of a theory of representation allows us to conceive a connectionist implementation of the validation of a representation by a process of selection of prerepresentations. Nevertheless, not all representations possess the statute of knowledge, the notion of which relates to higher-order social representations and the modalities of their evaluation, as we shall see.

The Raging Beast—Cognition and Language

Considerations of beauty and good discussed so far make up an inventory of thoughts about the interface between the science of man and the science of the brain. Using the principle of "top down" as we do in this book, we shall next tackle areas of neuroscience that since quite recently have stimulated active scientific research, both theoretical and experimental: those of consciousness and language. Only a few years ago it was inconceivable for a neuroscientist to use the word *consciousness* in serious scientific work, at the risk of losing the respect of his colleagues. Happily, this is no longer the case today: a real neuroscience of consciousness is spreading actively throughout the world. As to the science of language, it already underwent a revolution at the end of the nineteenth century, thanks notably to the work of Paul Broca, who localized where in the brain language came from. Since then it has benefited from new discoveries about the processes of cerebral plasticity and epigenesis that allow a better understanding of how the faculty of language, unique to man, has helped bring cultural diversity and individual identity to our society. Thus, Vico's beast of primitive humanity has become a civilized human being.

THE NEURAL BASIS OF CONSCIOUSNESS

The word *consciousness* is used by philosophers, moralists, and neurobiologists in very different ways, which it is important not to confuse. According to Henri Ey (1900–1977), *cum scientia,* from which the Latin *conscientia* was derived, includes "knowledge of the object by the subject" and, inversely, "reference by the object to the subject itself": an individual is both the subject of his knowledge and its author. For Kant unity of representations demanded unity of consciousness for their synthesis: consciousness is the highest point on which depends all use of knowledge. For Bertrand Russell (1872–1970), consciousness could also be seen as a relation of subject and object. For Martin Heidegger (1889–1976) it was appearance (*Erscheinung*). There is no limit to the number of definitions. Perhaps one way to begin is to restrict oneself to common usage, that of moral consciousness or conscience, the ability to judge what one ought to do, as well as reflection and emotion concerning what one has done.

For the neurobiologist consciousness relates to a function at the highest levels of organization of living beings. I discussed earlier the notion of levels of hierarchical organization or complexity in living organisms and particularly the presence of several such levels in the brain. This concept, which implies an objective relationship between structure and function, differs from that proposed by David Marr in 1982 and widely employed ever since. According to Marr, understanding a system for information processing necessitates the synthesis of partial solutions into a coherent whole at three descriptive levels: the computational (the aim of the calculation, its fitness for purpose, and the logic of its strategy); the algorithmic, which ensures the representation of the input-output relationship; and the implementational, used in computer hardware or neuronal networks. In fact, in the case of the brain, considered a complex system of information processing, the elaboration of such a general model could depend at each organizational level on Marr-style descriptive processing, which considers regulation between levels as well as global control processes.

The first evolutionists, particularly Lamarck and Spencer, recognized that the nervous system develops progressively from the simplest organisms to higher ones by gradual, undetectable developments. In his *Zoological Philosophy* Lamarck distinguished a "singular faculty with which certain animals and man himself are endowed," which he called "inner feeling." It received "emotions" from the intellect as well as from sensation or need, but it was located "below" volitional judgment by the "organ of intelligence" (the cerebral hemispheres). Spencer, in his *Principles of Psychology* (1855), proposed the evolution of nervous systems from simple to complex, whereby convergence and divergence of fibers ensure higher integrative relationships, "new grouped states" inserted between "primitive grouped states." The genesis of nervous organs accompanied by the genesis of corresponding functions, and consciousness independent of the immediate environment, thus became feasible. This concept seems to me to some extent the cerebral homologue of the *milieu intérieur,* defined at the same time by Claude Bernard for the organism as a whole.

At the end of the nineteenth century Hughlings Jackson, still somewhat misunderstood to this day, introduced the concept of evolutionism in mental disorders. According to his Croonian Lecture in 1884, disorders of the nervous system should be considered reversions of evolution, that is, dissolutions. The higher nervous centers form the physical basis of internal evolution. The dissolution caused by a lesion or disease "is a process of undevelopment . . . in the order from the least organised, the most complex, and the most voluntary, towards the most organised, most simple, and most automatic." Evolutionist theories, which are also those of modern neurobiologists, agree on the fact that consciousness emerges at the highest and most complex organizational level of the central nervous system, that it creates a relatively independent or autonomous milieu, a space for simulation and potential action, where internal evolution can develop, realizing a considerable economy of time, experience, and energy in the planning of an action in the outside world.

William James's *Principles of Psychology* (1890) remains a

seminal text in the psychology of consciousness, and it has lost none of its modernity. According to James, psychology is the description and the explanation of states of consciousness, and it has to be treated analytically, as a natural science. He even suggested the reduction of psychology to conscious experience and did not hesitate to write, "The immediate condition of a state of consciousness is an activity of some sort in the cerebral hemispheres." He tried to reconcile mentalist and physicalist viewpoints, not without difficulty, and finished, at the end of his life, by considering that it is not consciousness that should be considered by the psychologist, but only conscious experience. James never accepted the notion of an *unconscious* process.

To my mind, the following are his most pertinent contributions. He conceived of Me as a state of thought related to inalienable personal consciousness. There exist a Material Me, a Social Me, and a Spiritual Me, which are based on perceived states of consciousness. Consciousness is ever advancing, and intrinsically variable: "No two ideas are ever exactly the same." Thought is continuous, "without breach, crack, or division." Consciousness is like a bird, however, sometimes flying, sometimes not. The brain is an organ in unstable equilibrium and changes constantly. Personal consciousness selects certain elements of its content and rejects others, and we ignore most things that are before us. The problem is not what we want to do, but rather what we want to be. James's notion of the moral importance of effort is of interest to the modern cognitive psychologist. "The question of fact in the free-will controversy is thus extremely simple. It relates solely to the amount of effort of attention which we can at any time put forth."

In contrast, Pierre Janet, Jean-Martin Charcot's disciple and successor, wrote in *Psychological Automatism* (1889) that "all psychological laws seem false if one looks only for their application to conscious phenomena of which an individual is aware. Constantly one encounters events, hallucinations, or acts that seem inexplicable in these terms." So unconscious acts must exist. As to Freud, he generalized and systematized the notion of the unconscious, or subconscious, introduced by Josef Breuer (1842–1925), as an important component of psychic life. Separating a conscious from an unconscious domain, he considered consciousness as a "sense organ" used to perceive the extent of the conscious domain. Freud, in his study on hysteria with Breuer in 1895, emphasized that powerful mental processes remain hidden from consciousness: those representations of which we are aware are conscious, but, besides these, other thoughts exist that qualify as unconscious. Hysterics suffer particularly from abnormal states of consciousness, in which pathological representations appear. Freud saw in the subconscious the foundation of all psychic life and was gradually drawn into speculative philosophy, which sometimes led him to overestimate the role of the subconscious. In his *Project for a Scientific Psychology* (1895), not published during his lifetime, Freud proposed some important, but inaccurate, concepts—for instance, that the nervous system forms a continuous network, devoid of synapses or inhibition. From Santiago Ramón y Cajal we know that the nervous system is discontinuous: neurons are contiguous at synapses. Further, for Freud the nervous system was a passive receptacle for energy and information, not suited to creating them or releasing them other than by motor activity. Today we know the contrary: the nervous system is spontaneously active and functions in a projective fashion.

Comte denied the validity of introspection but emphasized the social dimension of consciousness. According to

John Watson (1878–1958), the founder of behaviorism, consciousness is nothing else but the "soul" of theology. Psychology is an objective experimental branch of natural science. Its theoretical aim is prediction and control of behavior. Introspection is not an essential part of its methods. Watson, however, founded a major new experimental field that favored the study of animal behavior removed from any anthropomorphic context.

In his work on conditioned reflexes, Ivan Pavlov (1849–1936) did not refer explicitly to the conscious or the unconscious, but he distinguished between higher brain centers of the cortex, to which he attributed characteristics of synthesis and analysis, and subcortical centers responsible for more fundamental activity of an organism such as feeding, sex, orientation, and aggression (part of what we have since come to think of as the limbic system). For Pavlov the cortex had a mainly inhibitory role on subcortical centers, and so he agreed with Hughlings Jackson. The cortex inhibits subcortical excitation and compulsive or affective activity, or automatisms. Seen this way, hysteria is a physiological disconnection between the cortex and subcortical centers so that the latter dominate the life of the hysteric by unnatural emotional influences. There is thus no explicit reference to consciousness, but rather to rationality for cortical activity and subjectivity for subcortical activity.

Though Ludwig Wittgenstein (1889–1951) objected to mentalist language as a general "disease of thought," neobehaviorism rehabilitated consciousness as a natural phenomenon. Logical behaviorism agreed with the views of neurobiologists in seeing mental states as systems for action with a power of causality. For D. M. Armstrong mental states were really only physical states. For Popper consciousness in higher organisms was the top hierarchical control system for the elimination of errors.

Functionalism attempted a synthesis of behaviorism and information cybernetics by playing down the role of the nervous system, although claiming to be materialistic. Taking inspiration from such varied sources as Descartes, Gall, Jean-Pierre Flourens, and Pavlov, Jerry Fodor (1983) distinguished two systems. Input systems are peripheral and specialized, or modular, and process information rapidly in an encapsulated and unconscious way, impenetrable to knowledge and higher brain function. Isotropic central systems are slow and accessible by consciousness, where "computational isotropy seemed to go hand in hand with neuronal isotropy [*sic*]."

In 1983 Philip Johnson-Laird envisaged the possibility of constructing automata with a degree of consciousness, which was no more than a property of a certain class of algorithm. He envisaged phylogeny of automata at three levels: a Cartesian, using no symbolism; a Craikian (after Kenneth Craik's concepts of 1943), which constructed symbolic models of the world in real time and possessed a rudimentary awareness, such as children or animals do; and, finally, self-reflective systems with the capacity for recursive embedding of models within models, and possessing a capacity for self-awareness, intentional behavior, and communication.

For the sake of completeness, we should mention the philosophical theories of phenomenology of Edmund Husserl (1859–1938) and Maurice Merleau-Ponty (1908–1961). For the latter phenomenology had "the ambition of philosophy as an exact science, but also an account of space and time in the real world. It is an attempt to describe our experience as it really is, without regard to psychological genesis and causal explanations." The most important contribution of phenomenology is perhaps its having united extreme subjectivism and extreme objectivism in its view of the world and rationality. It emphasizes the characteristics of consciousness that are relevant to a given subject that one could not dismiss with a stroke of the pen. It in no way opposed a materialistic concept of consciousness. As Hippolyte Taine (1828–1893) declared, subjective and objective comprise "a single and same event known under two aspects."

NEUROPSYCHOLOGY

The vocation of neuropsychology is to explore the consequences of lesions due to trauma, vascular accidents, or genetic disorders on higher brain function. In the case of vision, Bartolomeo Panizza in 1856 described central blindness after a lesion of the occipital region of the cerebral cortex. This observation was repeated by Hermann Munk in 1881 in dogs, which could still see and avoid objects after an occipital lesion, but could not recognize them. Hughlings Jackson noted that certain cortical lesions in humans lead to the phenomenon of imperception or asymbolism, which Freud called *agnosia:* elementary visual processes are preserved, but there is defective object recognition—Kant's *faculta signatrix.*

The phenomenon of blindsight depends on a different experimental paradigm. A lesion of the primary visual cortex causes a scotoma, a loss of elementary visual perception, in certain parts of the visual field. When a light is flashed in the blind field of a patient with a visual cortical lesion and therefore "blind," and he is asked to move his eyes (E. Pöppel et al., 1973) or point a finger toward the flash (L. Weiskrantz et al., 1974), he made the movement in the correct direction while still denying having ever seen the light: there is unconscious vision. It is due neither to diffusion of light in the eye nor to residual vision in the cortex because subjects succeeded in discriminating a black stimulus on a white background (to counter the diffusion problem), but failed to distinguish between a square and a rectangle, a task that necessitates cortical integrity. These heroic methods, which forced a subject to undertake an uncomfortable introspection, have recently been enriched by subtler techniques. One experiment has demonstrated spontaneous interaction between a stimulus presented in the blind visual field and another in the normal field by, for example, requiring the subject to complete a partial drawing. Another showed involuntary reflex responses to stimuli in the blind field, such as electrical conductivity in the skin or pupil diameter. Such experiments have shown that a lesion of the primary visual cortex is accompanied by dissociation between conscious and unconscious visual processes, and that access to a conscious level requires integrity of the primary visual cortex. The pathways involved in blindsight are secondary, that is, involving the LGN and secondary visual cortex, but also the superior colliculus of the midbrain and the thalamic pulvinar and their projections to the secondary visual cortex.

We know that the temporal cortex is involved in responses to complex stimuli in monkeys and humans, such as color combinations, abstract figures, hands, and faces. Bilateral lesions of these areas can cause selective loss of recognition of familiar faces, or prosopagnosia. Antonio Damasio showed, however, that certain patients who claimed not to be able to distinguish familiar faces manifested different changes in skin conductivity when exposed to familiar and unfamiliar faces. Once again we note dissociation between open or explicit recognition and unconscious discrimination between stimuli. Along similar lines, John Marshall and Peter Halligan in 1988 showed that patients who suffered from heminegligence of extracorporeal space could utilize information perceived in the neglected field. For example, flames alongside the image of a house on the neglected side were taken into account by the subject although he explicitly denied the existence of the flames. Overall, these studies showed an objective distinction between conscious and unconscious levels as well as the role of "classic" visual pathways in accessing the conscious ones.

The relationship between consciousness and time was a feature of William James's considerations. Consciousness is a dynamic entity. According to Henri Ey, a conscious being possesses a mastery of time through consciousness. The theoretical implications are important. It is not surprising that such a distinguished theoretical physicist as Roger Penrose should seize upon them. Addressing a cultivated, but hardly critical, audience, he did, however, go too far when he wrote in *The Emperor's New Mind* (1989) that "there could conceivably be some relation between this 'oneness' of consciousness and quantum parallelism" or that "single-quantum-sensitive neurons" play an important role in the brain. After suggesting that "a conscious 'mental state' might in some way be akin to a quantum state," forgetting the enormous difference in level of organization between the electron and the brain, Penrose came to the defense of the "anthropic principle which asserts that the nature of the universe that we find ourselves in is strongly constrained by the requirement that sentient beings like ourselves must actually be present to observe it." A self-confessed Platonist, he wrote: "Recall that Plato's world is itself timeless. The perception of Platonic truth carries no actual information . . . and there would be no actual contradiction involved if such a conscious perception were even to be propagated backwards in time!"

This explains Penrose's interest in the controversial experiments of the 1980s by Benjamin Libet. Initially he asked a subject to perform a simple motor task, like pressing a button, while the electrical activity from the subject's cerebral cortex (the electroencephalogram, or EEG) was recorded. There was a delay of some 200 milliseconds between the subject's deciding to make the movement and the actual movement, which was not surprising. But the EEG track of cortical activity showed activity in the secondary motor cortex (the readiness potential) some 300 milliseconds before the decision to move. This means that there was unconscious cortical activity before the subject decided to make a movement. In this case, all that one could do to stop the movement might be to inhibit this readiness potential. Later Libet worked with volunteers who needed to undergo brain surgery and compared the time course of their subjective perception of an electric shock to the skin compared with the EEG. There was an electrical response in the somatosensory cortex about 50 milliseconds after stimulating the skin, whereas conscious awareness of the stimulus took much longer, about 500 milliseconds, the time needed for "neural adequacy." We can react to a stimulus to the periphery, although unconsciously, in as little as about 50 milliseconds. Libet attempted to reproduce the skin stimulus by stimulating the somatosensory cortex directly. Indeed, he obtained a sensation, not necessarily identical to that obtained by stimulating the skin, but only if he continued sixty stimulations per second for up to 500 milliseconds, a period similar to the time needed to reach neural adequacy. Libet tried to compare the patient's subjective perception of the physiological stimulus and the minimal electrical stimulation of the cortex by stimulating the skin 200 milliseconds after the beginning of the cortical stimulus, and he noted that the subject perceived the skin stimulus before the cortical one. Libet explained his findings by suggesting that subjective conscious awareness of the stimulus was projected backward in time to the moment when it was experienced unconsciously in the cortex. In fact, Libet was comparing responses to very different stimuli. Access to consciousness was in one case physiological (stimulation of sensory receptors), and in the other it was artificial: the cerebral cortex is reputably difficult to stimulate, particularly because of the presence of multiple inhibitory circuits. Furthermore, long stimulus times were necessary before a subject reported any sensation at all. In another series of experiments Libet stimulated the skin 200 to 500 milliseconds before stimulating the cortex to which the region of the stimulated skin corresponded. He reported that the subsequent cortical stimulation blocked the effect of the skin stimulus: there was backward masking of the latter by the former. On the other hand, if the cortical stimulation was in the hemisphere opposite that to which the stimulated skin corresponded, masking did not occur: the subject perceived the skin stimulus as normal before the cortical stimulus.

The most plausible interpretation of these experiments, proposed by Patricia Churchland, is that when the two responses occur in the same hemisphere, the cortical stimulation deletes the memory of the skin stimulus and interferes with its full perception. This interpretation of Libet's results, if they indeed prove reproducible, supports the idea that conscious perception of a stimulus requires complex processing (as demonstrated by P300 waves), much longer than the preattentive analysis of the stimulus. There is no retrograde effect but, rather, an anterograde disturbance of a process of reconstruction of temporal evolution at the conscious level. There is thus subjective recomposition of real time at the conscious level, according to dynamics that are not necessarily the same, a concept that agrees with the ideas of William James.

The experiments of Paul Kolers and Michael von Grünau in 1976 can be interpreted along similar lines. With a tachistoscope, an instrument capable of producing very brief light stimuli, they produced on a white background two successive flashes composed of different colored squares 0.9° × 0.9°, 36° apart, each for 150 milliseconds and separated by 50 milliseconds. Subjects had an illusory perception of the first square moving toward the second. When asked to define the point where the color changed using a movable monochrome spot, they systematically proposed the middle of the interval between the two. The result was the same, irrespective of the paired colors or the subject. When subjects viewed a triangle and a square they perceived a continuous evolution of one figure to the other by rotation at the hypotenuse. So color change was abrupt, irrespective of the shape, but the change in shape was gradual. In both cases there was of course no going back in time from the second figure to the first, in either color or shape. The most plausible explanation is that already mentioned, of a deferred reconstruction of the sequence of events at a conscious level accompanied by an illusion of movement and reorganization in time of the sequence of the two visual stimuli. This view agrees with the multicopy model of consciousness suggested by Daniel Dennett, according to which multiple versions of an experience can be revised at great speed by addition, incorporation, amendment, or overwriting of the contents; no version is more correct than another. There would be a sort of flow or narrative sequence with a differential recomposition, producing sometimes abrupt, sometimes gradual perceptual changes. The results could also be interpreted simply on the basis of the model of a conscious global neuronal workspace, as we shall see in detail later.

In defining consciousness, we need an egocentric reference. For William James, "This Me is an empirical aggregate of things objectively known." For Henri Ey, to be conscious was "to dispose of a personal model of the world"; "the self stems from the autoconstruction of its own system of values derived from experience of life." Everyone's primordial experience is that of his own body, Henri Hécaen's (1912–1983) *somatognosia,* or body image. There are several pathological correlates, *asomatognosia,* often due to lesions of the inferior parietal lobe of the right hemisphere, in which a patient may ignore unilateral paralysis or have the impression of having only one half of his body. Other patients present with deficits in the designation of parts of their own bodies, often associated with frontal lesions, or in the orientation of their body in extrapersonal space, often after parietal lesions. In this context, phantom limbs are illusions of egocentric body consciousness that affect more than 90 percent of amputees, as described by Ambroise Paré (1510–1590). The phantom limb is initially described as having a definite shape, retaining the same sensations as the lost limb, and moving in space to seize objects or to walk. Gradually the phantom limb changes to adapt to the stump. Phantom limbs can appear even without amputation after anesthesia of the nerves to the limb. The origin of a phantom limb is obviously neural, but it might not be due simply to spontaneous activity in the somatosensory cortex. In 1968 Hécaen suggested that the existence of a body image, like phantom limbs, is related to "reafferences" (originally coined by Erich von Holst and Horst Mittelstaedt in 1950, and renamed reentry by Edelman), by which a corollary discharge, invoked by Hans-Lukas Teuber in 1951 and Richard Held in 1961, enabled the motor system to act retroactively on the sensory system. In this way invariant representations would be formed, characterized by an autonomy or constancy in relationship to the outside world and the body itself.

Many other bodily illusions or hallucinations have been described. Heautoscopy refers to a patient with a right hemispheric lesion who sees himself as if in a mirror, or may feel he is looking at himself from inside. There are auditory hallucinations due to problems in the temporal lobe, such as indistinct sensations of noise (tinnitus)—murmurs, the rustle of wind, the splashing of water, a train passing, or even a squeaking door, breaking crockery, clapping, or footsteps—that trigger great emotion and seem to come from an external source; the nineteenth-century Curé d'Ars, who often heard loud, violent noises, was one such case. There can also be melodies, music or words, phrases or even speeches, sometimes with a feeling of déjà vu. Similarly, there are reports of rudimentary visual hallucinations, such as spots of light, colored patches, parallel lines, spirals, or zigzags, associated with lesions of primary visual areas. More complex forms are objects, animals, religious visions of saints or burning bushes, and so on, which are related to the right-sided parieto-occipital cortex. With aging, diminution of hearing and vision is often accompanied by hallucinations. Finally, temporal lobe epilepsy can produce, between attacks, chronic changes in personality expressed as aggressiveness, disorders of sexual behavior such as transvestism or hypo- or hypersexuality, or events of a religious nature, like visions of heaven, angels, voices, ecstasy, or sudden conversion, that of Joseph Smith, founder of the Latter-Day Saint movement, being one famous case. There can be hypergraphia (writing obsessively for hours on end, marked by repetitive "litanies," perhaps with a moral or religious bent), or pedantic oration, described as unctuous, viscous, or even adhesive by Stephen Waxman and Norman Geschwind in 1974.

Consciousness of others can also be pathological. According to Uta Frith, this could be typical of autism. Hans Asperger (1938) and Léo Kanner (1943) introduced the term *autism* to describe a condition in infants characterized by narrowing of relations with persons in the outside world, abandoning of social structures, and withdrawal into the self. For Kanner the autistic child was characterized by its isolation from the world of people and a desire for immutability, and sometimes by small areas of exceptional ability for such things as memory tests, drawing, construction games, and mathematics. According to the International Classification of Diseases by the World Health Organization, autism is characterized by deficits in social interaction and communication, and restricted, repetitive behavior. Its incidence could be 2 to 3 per 1,000 births, and it affects 2.5 times more boys than girls. According to Uta Frith, *enfants sauvages* like Victor de l'Aveyron and Kaspar Hauser, the fools for Christ of old Russia, and even Sherlock Holmes would all be autistic. An intelligent robot, soulless and machinelike, devoid of any feeling for others, might have behavior comparable to that of an autistic. There is a variety of evidence for autism being biological in origin: there are anomalies in the EEG, anatomical abnormalities of the cerebellum, high incidence of a fragile X chromosome (in 20 percent of autistic children), and genetic mutations pertaining to synaptic development, as described by Thomas Bourgeron in 2007. Perinatal lesions (in 37 percent of autistic children) and viral infections have also been implicated. Although no single cause has been identified, these various factors could be related to common cognitive deficits. Several of the handicaps faced by autistic infants, such as loss of global vision and excessive attention to detail, repetitive activity, inflexibility, and distractibility, are reminiscent of adult frontal lobe syndromes. For Uta Frith, John Morton. and Alan Leslie in 1991, autistics have a deficit of the capacity, normally particularly well developed in humans, to predict and explain the behavior of other people in terms of their mental state, their beliefs, their intentions, and their emotions. Tests with comic strips enable such deficits to be measured. Autism seems to be a biological disorder that alters consciousness of self and others, probably involving the frontal cortex and the conscious global neuronal workspace.

Regulation of states of consciousness should be distinguished from that of the contents of conscious space. Rodolfo Llinás in particular, as we shall see, has studied the regulation of states of wakefulness and consciousness. Sleeping and dreaming represent states of consciousness distinct from that of wakefulness, the neural basis of which has been the subject of extensive research, notably by Michel Jouvet, summarized in 1999. In 1959 he described paradoxical sleep, when the individual is neither asleep nor awake but under the influence of a third state of the brain. This discovery came soon after the description of rapid eye movement (REM) sleep by Nathaniel Kleitman, William Dement, and Eugene Aserinsky. The importance of structures in the reticular formation of the brainstem in these states is indisputable. The pharmacology of waking and sleeping is very rich, and we shall return to it in more detail.

Contemporary Theories and Debates on Consciousness

THE SIGNIFICANCE OF THE THALAMOCORTICAL SYSTEM

Rodolfo Llinás and Denis Paré in 1991 based their studies of consciousness on electrophysiological recordings from the cortex, which revealed remarkable similarities in electrical activity during waking and REM sleep. Further, the threshold for sensory stimulation needed to wake a subject from sleep was much higher during the REM phase than during the other phases of sleep. Their hypothesis was that wakefulness is an intrinsic state, fundamentally similar to REM sleep but dependent on sensory input. REM sleep is a "modified attentive state in which attention is turned away from the sensory input, toward memories." In one case there is opening to the outside world, in the other withdrawal to an inner world. Oscillating activity in the system linking the thalamus, the group of subcortical nuclei through which sensory afferents are relayed to the cerebral cortex, to the cortex itself produces a global change in cerebral function that controls alternating cycles of waking and sleeping and the level or state of consciousness of the subject. The rich reciprocal connections between thalamus and cortex participate in the genesis of such oscillations, the various modes of which signal distinct states of consciousness. In the relay mode the EEG is desynchronized, as it is in waking or REM sleep. In the oscillating mode the EEG is synchronized, as it is in slow-wave sleep, the normal deep sleep. The relay mode is associated with tonic discharge of thalamic neurons, the oscillating mode with bursts of discharges and long periods of inhibition. Cholinergic (using acetylcholine as transmitter) neurons in the brainstem intervene in the change from one mode to the other, and sensory inputs on waking set the timing of internal rhythms, correlating spontaneous and evoked activity. Consciousness is an intrinsic property resulting from the expression of these systems in conditions of defined coherence. It assumes the reconstruction of outside reality into an inner neural reality, which also ensures temporal coherence throughout the brain. The thalamocortical system intervenes in the temporal unification of fragmented components of external reality and a person's inner being into a single

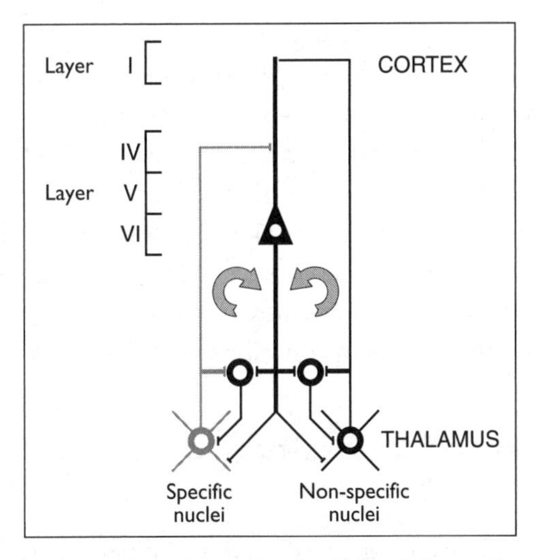

Figure 11: *Llinás and Ribary's hypothesis of oscillations between thalamus and cortex.* It was based on specific thalamic nuclei corresponding to layer IV of the cortex; nonspecific nuclei correspond to superficial layers, ensuring an overall regulation of cortical neurons. From U. Ribary, "Dynamics of Thalamo-cortical Network Oscillations and Human Perception," *Progress in Brain Research* 150 (2005): 127–142; reprinted with permission from Elsevier.

construct, the self. According to Llinás and Paré, subjectivity, the self, is born of dialogue between thalamus and cortex.

The thalamus contains several nuclei, groups of cells with different inputs and outputs. Anatomical and functional analysis of thalamocortical connections has revealed two main categories of nucleus (Figure 11). Specific nuclei receive inputs from ascending sensory systems and correspond to layer IV of the cerebral cortex: they are organized topologically in both thalamus and cortex, so they help produce maps of cortical representations of the body. Besides these, we distinguish nonspecific, intralaminar nuclei in the thalamus, which correspond to superficial cortical layers I, II, and III. They are found outside the specific thalamic nuclei and correspond to the cortex more widely, over several cortical areas. They might participate in overall control of states of consciousness within the cortex, and therefore to a conscious neuronal workspace.

ELECTROPHYSIOLOGY, EEG, AND EVOKED POTENTIALS

The analogy between states of waking and paradoxical or REM sleep has often been emphasized in the past. At the transition from waking to slow-wave sleep, fast waves become slow and synchronized, whereas the transition from slow-wave to REM sleep is marked by a return of fast, desynchronized waves, a common feature of waking and paradoxical sleep (Figure 12). In contrast, sensory inputs have very different effects in waking and REM sleep. During REM sleep there is desynchronization and muscle atonia, and the threshold for sensory inputs is much higher than during waking. Thus, there is a sort of isolation of paradoxical activity. As I said above, for Llinás and Paré REM sleep is a modified attentive state for which attention is turned from sensory inputs toward internal memory. Conversely, waking is similar to REM sleep, but dependent on sensory input.

What is the origin of thalamocortical oscillations, especially fast ones at 40 hertz? In vivo and in vitro recordings of thalamic neurons have shown that a progressive depolarization of the membrane of these neurons leads to a gradual change

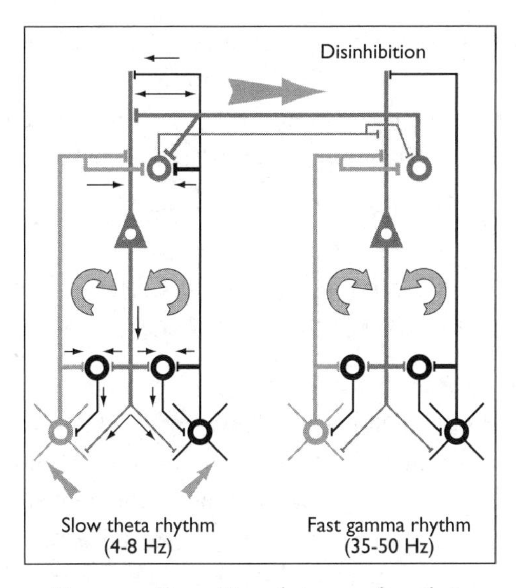

Figure 12: *Transitions between slow theta waves of sleep and fast gamma waves of waking.* On the right, reduced inhibition by GABAergic neurons (in black) causes disinhibition in the system. From R. Llinás and M. Steriade, "Bursting of Thalamic Neurons and States of Vigilance," *Journal of Neurophysiology* 95, no. 6 (2006); reprinted with permission from the American Physiological Society.

from a series of isolated action potentials to high-frequency bursts. Mircea Steriade's recordings in vivo of cortical neurons in the cat showed that depolarization of these neurons progresses from brief bursts at slow frequency, corresponding to slow-wave sleep, to maintained continuous high-frequency discharge, corresponding to wakefulness. This range of activity is controlled by neurotransmitters or neuromodulators. According to Steriade, excitation of muscarinic receptors by acetylcholine causes the opening of "inhibitory" potassium channels on inhibitory neurons and thus the global excitatory change from bursts to the continuous high-frequency rhythm found during waking. General anesthetics might block this modulatory effect indirectly.

THALAMOCORTICAL RESONANCE AS A NEURAL BASIS OF CONSCIOUSNESS

Llinás and Paré studied the transitions between states of consciousness by magnetoencephalography (MEG), which allows the recording of cerebral activity on a time frame similar to that of the EEG (Figure 13). In awake subjects coherent signals at 35 to 45 hertz were recorded, and the signal-to-noise ratio was very high. If the response to an auditory stimulus was then recorded in awake and sleeping subjects, an auditory stimulus led to a considerable increase of 40 hertz oscillation during waking, but no resetting, and therefore resonant amplification, during slow-wave or REM sleep. For the authors this meant that there is resetting during waking, the context being created by the brain, whereas in REM sleep sensory activity has no access to the mechanism that engenders conscious experience. The temporal coincidence between specific and nonspecific thalamocortical systems would provide access to conscious space.

CRICK, EDELMAN, AND BAARS: THE FIRST DEBATES ON NEURONAL CORRELATES OF CONSCIOUSNESS

In 1990 Francis Crick and Christof Koch adopted a simple proposal. In their opinion, the function of consciousness is "to present the result of various underlying computations and . . .

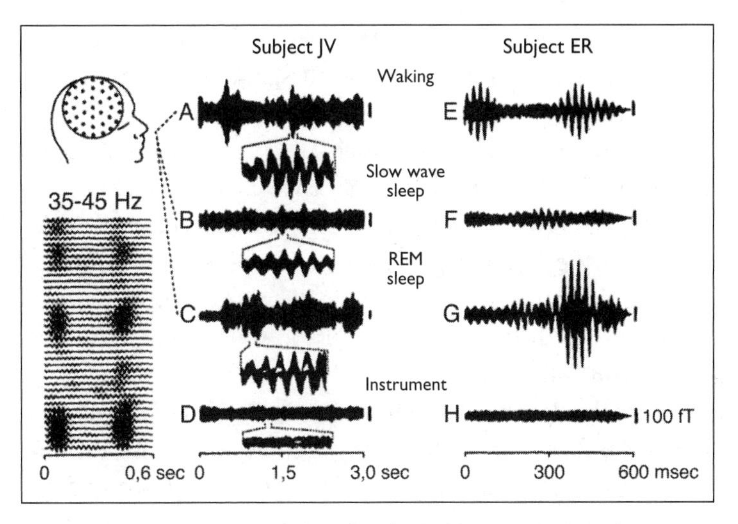

Figure 13: *MEG recordings of 40 hertz gamma activity during waking, slow-wave sleep, and REM sleep.* From D. Paré and R. Llinás, "Conscious and Pre-conscious Processes as Seen from the Standpoint of Sleep-Waking Cycle Neurophysiology," *Neuropsychologia* 33, no. 9 (1995): 1155–1168, reprinted with permission from Elsevier.

this involves an attentional mechanism that temporarily binds the relevant neurons together by synchronizing their spikes in 40 Hz oscillations." The binding between neurons required for coherence of consciousness is achieved by oscillations at 40 hertz, corresponding to synchronous discharges that placed the mental object in working memory. They based their idea on the observations of Charles Gray and Wolf Singer in 1989, but they seemed to identify coherence with oscillation, whereas, depending on the system involved, they are separate. Furthermore, the somewhat neglected results of Margaret Livingstone and David Hubel from 1981, using recordings from single neurons responding to visual stimuli during the transition from slow-wave sleep to waking, demonstrated that waking coincides with desynchronization of bursts of slow waves and a substantial increase in the signal-to-noise ratio. Contrary to the hypothesis of Crick and Koch, waking is accompanied by desynchronization of oscillations. Their ideas have been criticized on the grounds that oscillations do not necessarily coincide with the binding of neurons, and binding can happen without reaching the conscious workspace. Once again, it is important to distinguish between the *state* of consciousness and its *contents.*

In *The Remembered Present* (1989) Edelman suggested a biological theory of consciousness that developed the ideas he had earlier proposed in *Neural Darwinism* and *Topobiology,* but with new hypotheses. His basic hypothesis of reentrant signaling, put forward in 1978 as an essential part of the mechanism of neural selection, was actually suggested by me in *Neuronal Man* as possibly contributing to the establishment of a global state of consciousness. This idea offered a mechanism that enabled consciousness to emerge from the comparison of the activity of two sorts of neural system: one related to memory and constituting self, the other concerned with sensory interaction with the outside world and constituting non-self. The former includes the hypothalamus and pituitary, the brainstem, the amygdala, the hippocampus, and the limbic system in general. It becomes organized during development, in particular through the attachment of labels of values to perceptual categories. The other includes the cerebral cortex, thalamus, and cerebellum. A major input loop to the non-self system enables the formation of primary consciousness. Higher-order consciousness emerges with the development of categories related to the concept of self and with the acquisition of language. This model offered original features, but it was nevertheless somewhat reductionist in the assignment of self and non-self to such gross anatomical subdivisions of the brain. Further, it did not take into account the dynamics of consciousness and was not formulated in precise computational terms.

In *A Cognitive Theory of Consciousness* (1988) Bernard Baars dealt with the problem of consciousness on the basis of cognitive psychology and so updated the ideas of William James and Wilhelm Wundt. As a cognitive psychologist, Baars revisited the Kantian hypothesis of transcendental consciousness, with its capacity for global synthesis, as well as Fodor's central systems that he termed global workspace and that he contrasted with systems of automatic unconscious processors. In turn, the workspace broadcasts information from a system of processors to all other processors in such a way that the outcome can be compared to a "theater, a screen, a blackboard" (Figure 14). This metaphor is useful, for it includes consciousness of one thing at a time. At any given time many more things happen than we know about, and unknown events can happen backstage and control what penetrates to our subjective mind. Processors compete or cooperate so that their messages arrive in the global workspace. The global message of the various processors has to be coherent and informative, and thus adapted to the unconscious context. Unconscious context and conscious content interact to create a current of consciousness, as Henri Bergson (1859–1941) suggested in *Creative Evolution* (1998), and voluntary activity can be seen as a particular way to resolve a problem. Self can be considered a context dominating conscious experience, bringing information about "self as a context." Baars developed a series of successive models. His Model 1 simply contained competing input processors, conscious global workspace, and

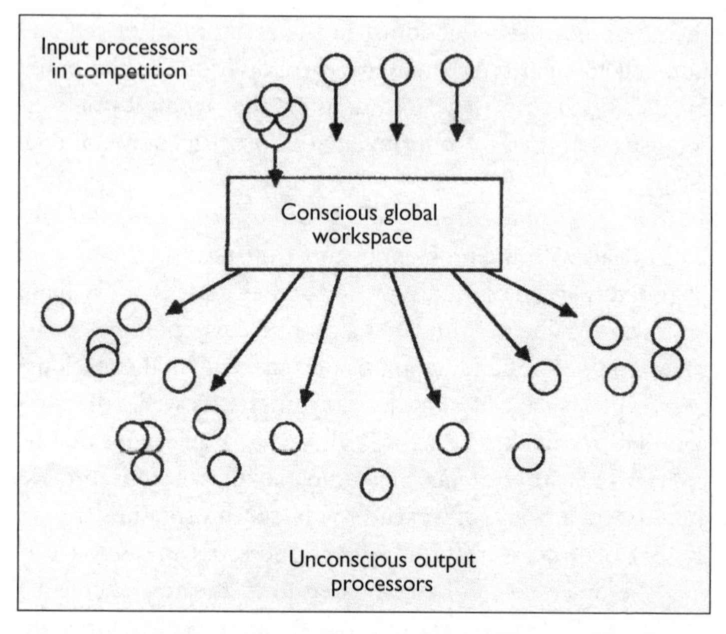

Figure 14: *Baars's model of a global workspace.* Baars proposed that the nervous system could be considered a collection of specialized unconscious processors, the interaction, coordination, and control of which require a central exchange of information in a conscious global workspace. The neural bases of the global workspace include the brainstem reticular formation and diffuse projections from the thalamus to the cortex. From B. Baars, *A Cognitive Theory of Consciousness* (New York: Cambridge University Press, 1988); reprinted with permission of Cambridge University Press.

unconscious output processors, along with possible feedback of inputs and outputs. Further models involved unconscious contextual systems, the role of attention and its voluntary control, and ultimately the supervisory context of self. So Baars's theory was based on a conscious global workspace, with specialized unconscious processors as perceptual analyzers, output systems, systems for action, syntax, planning, control, and goals. One of the objections to Baars's models is that they do not mention reward systems or selection on the basis of "value."

Shortly after, in 1991, Stanislas Dehaene and I elaborated the idea of a formal organism that successfully passed the Wisconsin card-sorting task. It had external and internal reward mechanisms, offered a neuronal mechanism for selection of clusters of rule-coding neurons, and possessed dynamics for evolution of rule-coding neurons related to the external or internal world. These elements served as a basis for the development of a more general model of the global neuronal workspace, including access to consciousness, by Dehaene, Michael Kerszberg, and myself in 1998. The identification of the neural basis of consciousness represents a genuine scientific challenge. Neurobiological data are accumulating on the distinction between conscious and unconscious compartments, and objective criteria for this distinction can be defined in order to elaborate a coherent computer model for access to consciousness.

THE DEBATE CONTINUES

In 2000 John Searle declared that the time had arrived to study consciousness like any other biological phenomenon. The problem was to discover which cerebral processes were responsible for the conscious state and how this state was explicable in terms of brain structure. In other words, what are the neurobiological correlates of consciousness? For Searle consciousness consists of a qualitative, subjective, unitary internal state. How can one define "in the first person" the cerebral mechanisms, which are objective chemical and electrical biological processes likely to produce subjective states of sensation and thought "in the third person"? Searle distinguished two things. First is the building-block approach, which consists of identifying the component parts, collecting them, and deducing the field of consciousness. Second is the unified field, according to which there are no elementary components, only modifications of an existing field of qualitative subjectivity. According to him, Crick's model belongs to the first approach, and Edelman's to the second.

Chris Frith and his colleagues tried to define experimental conditions for revealing the neuronal correlates of conscious experience. They attempted to show how to identify the distributions of neural activity specifically associated with conscious activity. They distinguished the level or state of consciousness, such as sleep or waking, that involves the reticular formation, the locus coeruleus of the brainstem, or the intralaminar nuclei of the thalamus. They also distinguished the contents of subjective experience, that of which we are conscious, and also the percepts, such as memory and attention, that involve specific areas of the cortex. We can access the contents of conscious experience only through a subject's own account. Frith emphasized the difference between an account of an experience and the conscious experience itself. An account is usually verbal, expressed through language to another person. Besides the verbal account, there exists a behavioral relationship of gestures or movements in the case of cognitive experimentation—for example, when one has to press a button to acknowledge the perception of an event. Research in this field implies correspondence between conscious experience and neural events recorded by electrophysiology or imagery. Frith and his colleagues distinguished at least three types of neural activity that has to be resolved in space and time: activities associated with conscious mental representations, those associated with sensory stimulation, and those associated with behavior.

Crick and Koch studied the visual system and tried to relate it to consciousness, using experiments on monkeys involving binocular rivalry and the hierarchy of visual pathways. Classically, the dorsal pathway is fast, in egocentric coordinates, and propagates online unconscious representations, such as eye or hand movement. The slow ventral pathway, in allocentric coordinates, propagates explicit representations with access to consciousness. It gives a better interpretation of the visual scene and contributes to the availability of motor systems. They proposed that neurons in layers V and VI of the cortex provide access to consciousness on the basis of oscillating thalamocortical properties. Activity of primary visual cortical neurons, without projections to the frontal cortex, however, would not be correlated with what we see consciously. What accesses consciousness is neural activity in the higher visual cortex with direct projections to the prefrontal cortex. We now know that the primary visual cortex can be mobilized by the conscious neuronal workspace.

Subsequently, Crick and Koch recalled reading Ray Jackendoff's 1987 book *Consciousness and the Computational Mind.* A linguist, a musician, a protégé of Chomsky, and an expert in informatics, Jackendoff placed consciousness at an intermediate level of representation, between a peripheral level of sensation and a central level of thought. We are not conscious of either the data from sensations or the form of thoughts. Jackendoff distinguished the phenomenological mind, the seat of experience of the world and our inner life, inaccessible to others, and the computational mind, which processes information and is a center for comprehension, knowledge, reasoning, and intelligence. Phenomenological and computational mind are two different levels of description of the physical body, the former constituting an abstract specification of the functional organization of the nervous system, to use computer programming terms. It is a mathematical model of the brain at work. For Jackendoff no activity of the computational mind is conscious. For example, when we think in words, the thoughts come in a grammatical form, with subject, verb, and object all in place and not the least perception of how the structure of the phrase is produced. He claimed that we hear an "inner voice" speaking in words. For vision, do only two and a half dimensions reach consciousness, as David Marr suggested, the third dimension remaining unconscious? Crick and Koch curiously added the notion of the homunculus, a little man in our head who perceives the world with his senses, thinks, plans, and executes voluntary actions. For them the homunculus is unconscious. Whatever the case, reading Jackendoff can be useful for trying to define what types of representation effectively reach consciousness. The question of a conscious detachment of the examined object and that of the representation of two and a half or three dimensions is important even if the idea of the unconscious homunculus seems difficult to accept. Finally, the prefrontal cortex cannot be considered simply the seat of high-level unconscious computations, as Crick and Koch proposed, in contradiction with their earlier concept.

After the first works of Edelman in 1989 on primary and higher-order consciousness, in 2000 he and Giulio Tononi turned to more elaborate modeling, yet often with no obvious relation to consciousness. In particular, they devised models of integration in the visual cortex and a thalamocortical model. Dealing with interconnections between multiple visual areas, each with a specialized function, such as movement, color, or shape, they proposed a model that resolved the problem of connectivity, which would involve not only reciprocal connections between areas but also evaluation systems, as well as control of eye movements. Based on empirical data on thalamocortical relations, this model was surprisingly related by its authors to access not consciousness, but visual processing! Then they approached the issue of the complexity of the nervous system. Edelman and Tononi questioned the validity of applying standard information theory to the nervous system. Information theory requires an intelligent external observer to code and decode messages with an alphabet of symbols. The approach they proposed was purely statistical and made no reference to an outside observer. They distinguished effective information, the number and probability of the states of the system that make a difference *within* the system itself, and mutual information, which measures the independence between two subsets of elements by bipartition of a given isolated system. For them the complexity, or total integrated information, corresponds to the sum of values of mutual information for all the bipartitions of the system. This complexity can vary with neuroanatomical organization. It is relatively low when connections are distributed independently, and maximal when one is dealing with a precisely connected domain of defined neurons. The greater the mutual information between each subset and the rest of the system, the greater the complexity. So the complexity of living organisms is situated "between crystal and smoke," as evoked by Henri Atlan some years ago.

Further, their dynamic core hypothesis took up the notion of integration with the idea that a system is integrated if its elements interact more strongly among themselves than with the rest of the system. There is internal cohesion and external isolation. For Edelman and Tononi clusters of neurons can contribute directly to conscious experience if they belong to a distributed functional cluster, which by reentry interactions in the thalamocortical system realizes a high level of integration in a few hundred milliseconds. There is a functional border between this cluster and the rest of the brain; it forms a dynamic core with clear functional boundaries. This dynamic core has properties of integration, a constantly changing composition, and a variable spatial distribution not localizable to a single place in the brain. It is neither coextensive with the whole of the brain nor limited to a subset of neurons, and it certainly does not correspond to an invariant group of cortical areas, for the same group of neurons can be part of the core at one time and be outside it at others. Thus, the precise composition of the core varies significantly from time to time for an individual and from one individual to another. It is at any given time unified, private, and divided. The definition of neuronal correlates of the dynamic core poses a problem because of the lack of a precise architecture for it in the brain and of a definition of any intrinsic properties of its neurons. We know only that there are long-distance connections between different parts of the brain, and these can differ from one brain territory to another. Edelman and Tononi tested their hypothesis in experiments on binocular rivalry using magnetoencephalography, but the variability of the data from these experiments posed a serious problem. Their model drew certain criticism. They did not distinguish between state and contents of consciousness, and they were firmly opposed to the idea of a definite neuronal architecture for consciousness.

Herbert Jasper (1906–1999) had for a long time taken a firm stand against such hypotheses. For him a specific system of neurons in the brain is involved in consciousness, including in particular the reticular systems of the brainstem and thalamus. There exists a neural architecture for consciousness. Epilepsy causes loss of consciousness, and so there are specific activities related to consciousness. Further, attention is an important aspect of consciousness, and it needs further analysis. For Jasper consciousness is a brain function involving and integrating multiple cerebral regions.

Additionally, Leonardo Bianchi, in his publication of 1921 on the mechanism of the brain and function of the frontal lobes, reported work he had carried out on dogs between 1881 and 1894. For him consciousness is not a faculty but an active part of psychic processes in a highly evolved brain. It is variable, and its evolution is endless and limitless. It progresses with the development of complexity in organisms, particularly that of the nervous system. Consciousness attains its apogee through deliberation, which depends on judgment, which is the result of impulses and inhibitions. The dawning of higher consciousness coincided with the evolution of the frontal lobes, the cerebral organ that founds, transforms, summarizes, and regulates the immense mental patrimony prepared by more posterior parts of the brain. The prefrontal areas relate to sociability and intervene in the great mental syntheses, an idea with which I would agree.

OUR FORMAL MODEL

In a deliberately Cartesian undertaking, in the sense that theory preceded it and experimentation accompanied it, we elaborated our formal model, mentioned several times already, of a conscious global neuronal workspace. It comprises a minimal, coherent, autonomous theoretical representation and has a mathematical form; but, being based on defined biological premises, it can lead to precise experimental predictions. This undertaking is not simply reductionist. Certainly it exploits knowledge of elementary structures, but it is completed by a critical step of reconstruction of these basic elements. The general hypothesis is that living organisms evolve by multiple assembly mechanisms and organizational selection adapted to environmental conditions. This capacity for assembly, or "tinkering," as François Jacob called it, involves a fundamental property of auto-organization from the most basic levels, as molecular assemblies amplify to form supramolecular, cellular, tissular, and organic structures, accompanied by ongoing contributions from variation and selection.

We used allosteric mechanisms to account for not only properties of receptors of neurotransmitters, but also the capacity for integration of multiple signals by these same receptors. In 1949 Donald Hebb proposed that if a synapse from an axon excites a postsynaptic neuron sufficiently to fire it, then that synapse will be reinforced in some way. It is possible to elaborate a Hebbian chemical synapse from elementary properties of allosteric receptors and to adapt the efficacy of a synapse by an adjacent synapse. Thus, a synaptic triad can be constructed that can memorize and recognize a temporal sequence of impulses transmitted by two synapses on a single-cell body (Figures 15, 16). From these triads we elaborated layered structures that could account, in a Darwinian process of trial and error, for the acquisition and production of a melody in which each note has a defined value in a sequence, which means recognizing a basic dependence on context.

In 1989 and 1991 Dehaene and I elaborated very simple formal models of "organisms" capable of learning a cognitive task with selection by reward. The delayed-response task includes "matching to sample" (task A, not B). The formal organism comprises at a minimum two structural levels: a basic sensorimotor level, at which there are modifiable synapses with the capacities of perception and prehension, and rule-coding clusters, composed of richly interlinked excitatory

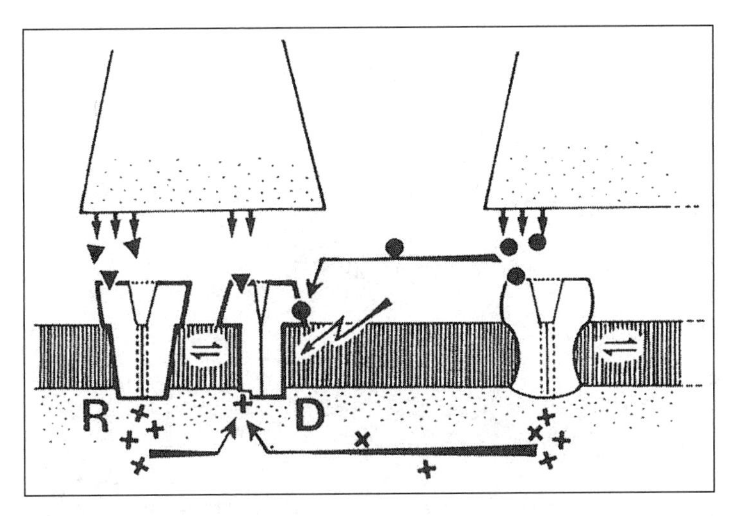

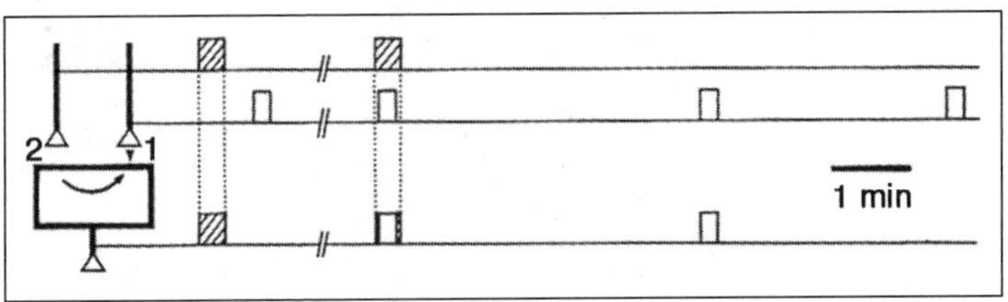

Figure 15: *Allosteric receptors and learning. Top:* a hypothetical model of control of synaptic efficiency that is based on allosteric transitions in the nicotinic receptor. Activity of the synapse on the right controls the efficiency of the synapse on the left at the postsynaptic receptor. Temporal coincidence between conditioned and unconditioned signals is read at the level of the postsynaptic receptor. *Bottom:* computer simulation of this mechanism is compatible with classic conditioning. From T. Heidmann and J.-P. Changeux, "Un modèle moléculaire de régulation d'efficacité d'une synapse chimique au niveau post-synaptique," *Comptes Rendus de l'Académie des Sciences* 295 (1982): 665–670; reprinted with permission.

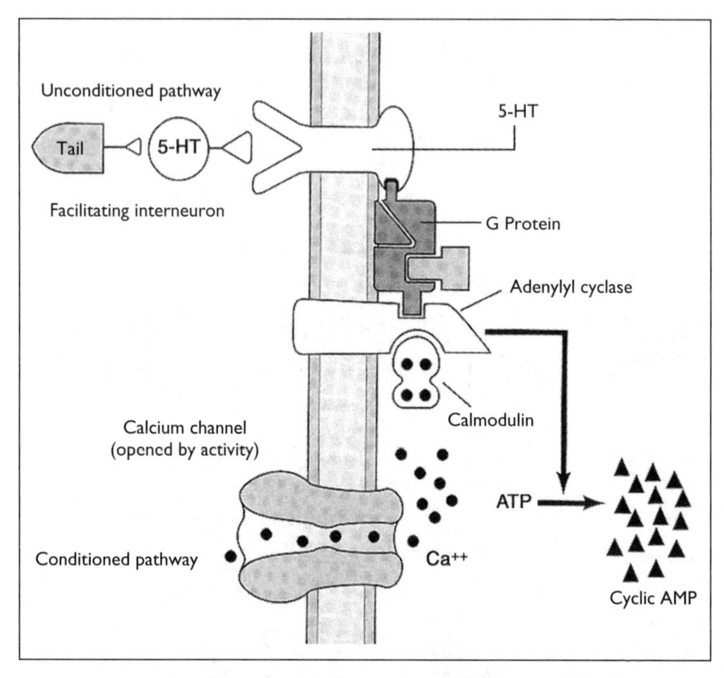

Figure 16: *Kandel's molecular model for classic conditioning in Aplysia.* Compare this to Figure 15. From E. R. Kandel et al., *Principles of Neural Science* (New York: McGraw-Hill, 2000); © The McGraw-Hill Companies, Inc.

neurons, which are mutually inhibitory by long axonal connections so that only one rule is active at a given time. The rule-coding neurons determine behavioral activities that, if successful, stimulate a positive-reward mechanism, which in turn stabilizes the particular rule-coding cluster active at that moment. So there is selection by reward. If the behavior results in failure, the consequence is destabilization of the rule-coding cluster and oscillation from one group of neurons to another until a new rule results in a positive reward. This model proposes that the reward system acts directly or individually on the rule-coding neurons by changing synaptic efficacy at, for example, allosteric receptors. A diversity generator enables the organism to select rules by trial and error. One can now speak of mental Darwinism.

The Wisconsin card-sorting task is more complex and is used to detect damage to the prefrontal cortex in patients. It consists of an improved delayed-response task using cards with symbols that differ in color, number, or shape. The patient is asked to classify the cards according to four key cards presented to him, following a rule of color, number, or shape. The patient must give the maximum number of correct answers and the experimenter replies that each is good or bad. Suddenly, the rule changes from, for example, a color rule to a shape rule. The patient must then note the change and discover the new rule. Results for this test are abnormal after prefrontal lesions, perseverance in error being a typical outcome. I propose a more complex structural basis than that of the deferred-response task, one that includes groups of neurons with motor intentions that can be active without being in action, as well as a self-evaluation circuit that enables us to evaluate tacitly an intention in light of already memorized rules. The formal organism so constructed succeeds in the Wisconsin card-sorting task. It has episodic memory and can reason. In fact, rules can be eliminated a priori by evaluation *before* results occur. So there can be a tacit internal test of a potential rule. This is already an attribute of consciousness.

The Tower of London test is even more complex, in that the subject is asked to pass from one configuration of beads strung on pegs to another configuration of the same beads. We propose a structure that incorporates a system of descending planning and of ascending evaluation by reward, which permits a hierarchical sequence of organized operations to attain a goal (Changeux and Dehaene, 1998).

THE GLOBAL NEURONAL WORKSPACE

What is the neuronal basis for making a conscious effort, such as an arithmetic calculation? A more general problem is that of tasks of mental synthesis, to solve a problem, involving several distinct modalities. There is no question here of the *state* of consciousness: the subject is conscious and awake. The problem is to know the *contents* of consciousness, or the conscious operations. The proposed structure (Figure 17) adopts earlier schemas with two principal levels and generalizes them. First is a neuronal workspace, which corresponds to Tim Shallice's supervisory attentional system, Jerry Fodor's central system, and Bernard Baars's global workspace. What distinguishes our model from previous ones is that we propose a precise neuronal architecture for our conscious workspace (Dehaene, Kerszberg, and Changeux, 1998). It would mobilize enormous pools of interconnected neurons with long axons and would involve several cortical areas (Figure 18). There would be dynamic recruitment with global integration of representations

Figure 17 (opposite): *Our model of a conscious neuronal workspace.* It involves a formal neuronal network of specialized modular processors, including sensory and motor systems, long-term memory, attention and evaluation systems, and a global workspace like the one Baars described, but with a neuronal basis of cortical interconnections via long-axon neurons. From S. Dehaene et al., "A Neuronal Model of a Global Workspace in Effortful Cognitive Tasks," *Proc. Natl. Acad. Sci. USA* 95 (1998): 14529–14534; copyright © 1998 National Academy of Sciences, U.S.A.

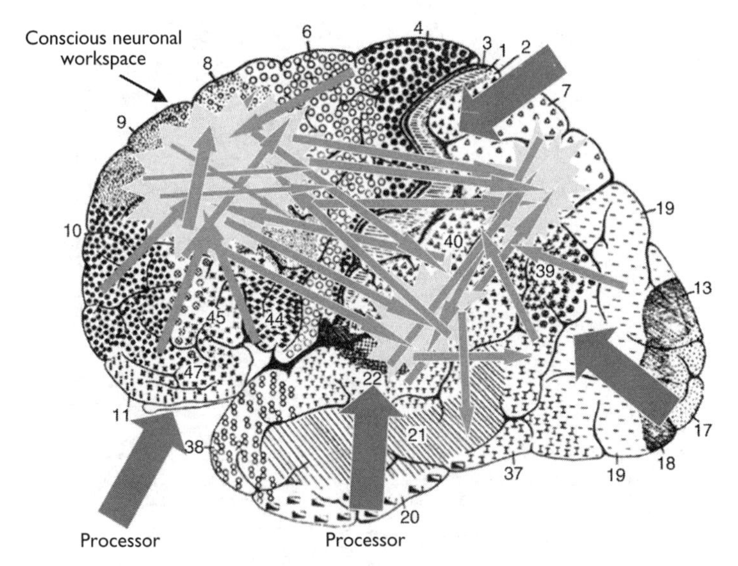
Conscious neuronal workspace
Processor
Processor
1
2
3
4
6
7
8
9
10
11
13
17
18
19
19
20
21
22
37
38
39
40
44
45
47

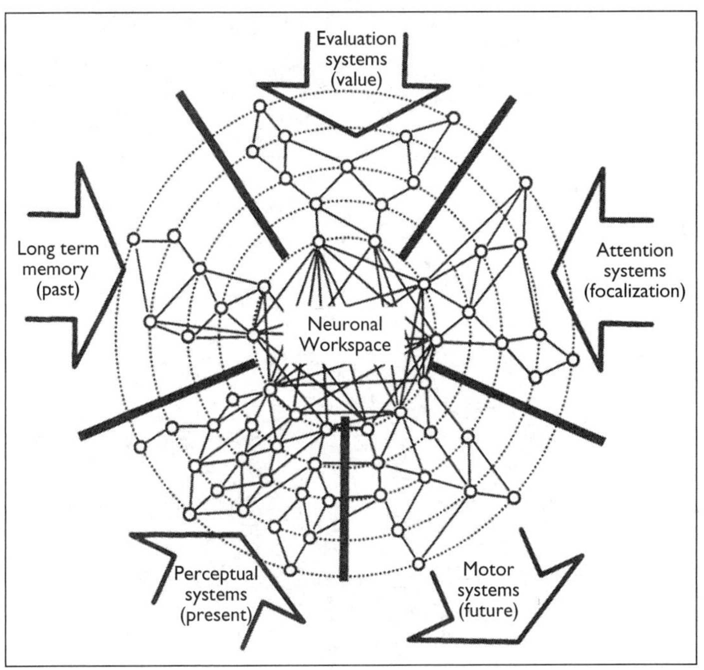
Evaluation systems (value)
Long term memory (past)
Attention systems (focalization)
Neuronal Workspace
Perceptual systems (present)
Motor systems (future)

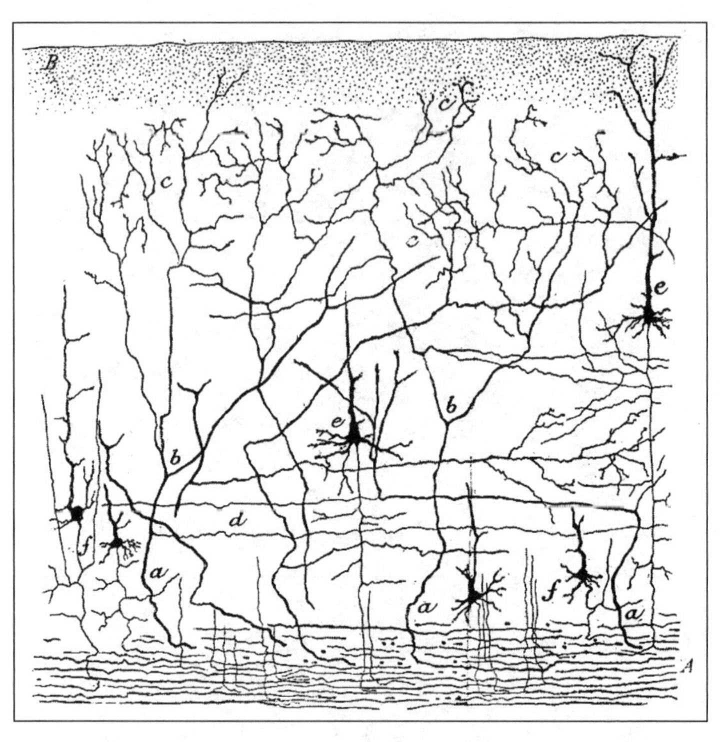
B
c
b
d
e
f
a
A

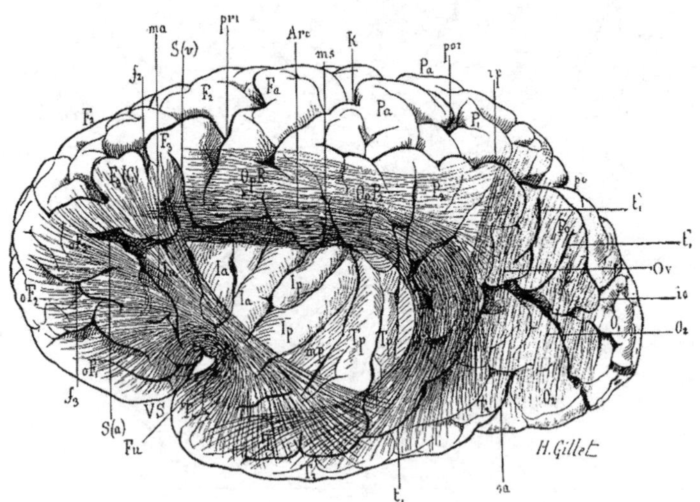
ma
S(v)
Art
ms
K
Pa
Fa
Ia
Ip
Tp
VS
S(a)
Fu
Ov
H. Gillet

possessing the properties of unity and diversity, of variability and competition, proposed by Edelman and Tononi, but in a physical domain, the neural structure of which is defined and limited. To this workspace is added an array of compartmented processors consisting of neurons connected by short axons, intervening in processes like vision, semantics, and motility. The novel anatomical hypothesis that is developed assigns a primordial importance to long-axon neurons, particularly abundant in layers II and III of the cortex, notably in the prefrontal and inferior parietal areas. Thus, we see the critical importance of the frontal lobes in consciousness. In the Stroop test a subject is asked to name the color in which the name of a color is written, such as the word *red* written in blue. The result is that usually the sense of the word is reported almost automatically, whatever color it is written in; the subject must make a real effort to correct himself. To do that, he uses the neurons of his workspace, which by trial and error check from top down the information provided by processors working from bottom up. Computer simulation of the model enables us to realize the selection dynamics of a global representation, and even to predict the dynamics of cerebral imaging that can be seen during the execution of this test (Figures 19, 20, 21). Abundant experimental evidence has been gathered by several groups since 1998 that agrees with this model of the global neuronal workspace.

Figure 18 (opposite): *Drawings of neuronal fibers in the cerebral cortex. Top:* Ramón y Cajal's drawing of long-axon neurons in the cortex. *Bottom:* Dejerine's figure of association fibers in the white matter of the lateral aspect of the left hemisphere. From J. Dejerine, *Anatomie des centres nerveux.*

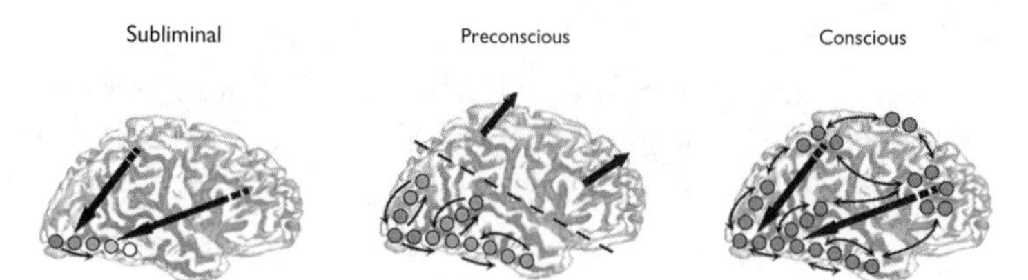

Figure 19: *Theory of the conscious neuronal workspace.* Diagram of the processing of subliminal, preconscious, and conscious visual stimuli, illustrating the mobilization of circuits of the neuronal workspace. From S. Dehaene et al., "Conscious, Preconscious, and Subliminal Processing: A Testable Taxonomy," *Trends in Cognitive Sciences* 10 (2006): 204–211; copyright © 2006, reprinted with permission from Elsevier.

Consciousness and Social Interaction

THEORIES OF COMMUNICATION: CODE MODEL OR INFERENTIAL MODEL?

Humans do not communicate directly from brain to brain, but by means of specialized communication processes, of which there are two principal theories. From the classical Aristotelian view to modern semiotics, the most commonly adopted theory involves codes. Communication means coding and decoding messages. Claude Shannon and Warren Weaver in 1948 proposed that messages containing meaning are transmitted as coded signals produced by one partner and received by another. They are propagated from transmitter to receiver by a physical channel, perhaps air or an electrical cable. The message, such as a series of letters forming a text, can be coded by electrical signals representing each letter. Communication requires that the same code be used throughout. In addition, it is susceptible to noise and errors, however minute. Internet addresses are an obvious example. This code model has been

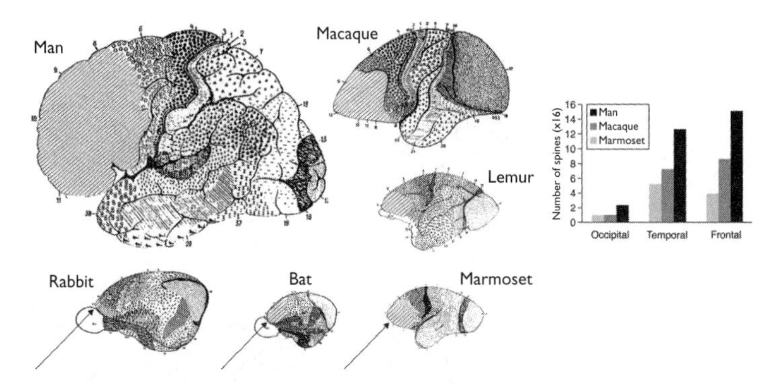

Figure 20: *Growth of the conscious workspace during evolution.* There has been an increase in relative surface area of the frontal cortex. From G. N. Elston, "Cortex, Cognition and the Cell: New Insights into the Pyramidal Neuron and Prefrontal Function," *Cerebral Cortex* 13 (2003): 1124–1138; reprinted with permission from the Oxford University Press.

widely adopted for theories of communication of verbal language, from Aristotle (via the *Port-Royal Grammar*, published in 1660 by Antoine Arnauld and Claude Lancelot), to Ferdinand de Saussure (1857–1913) and Lev Vygotsky (1896–1934). Codes link thought with sounds according to associative mechanisms that conform to empiricist theory. Although this model accounts for communication of thoughts, however, it does not explain their comprehension.

The inferential model developed by Paul Grice in 1957 and Dan Sperber and Deidre Wilson in 1986 was based on the idea that human linguistic communication cannot be reduced to an autonomous series of coded and decoded words. Words can be understood only in the context of the communicator's view of the world. Mutual knowledge is indispensable for communication. Effective communication takes place in a rel-

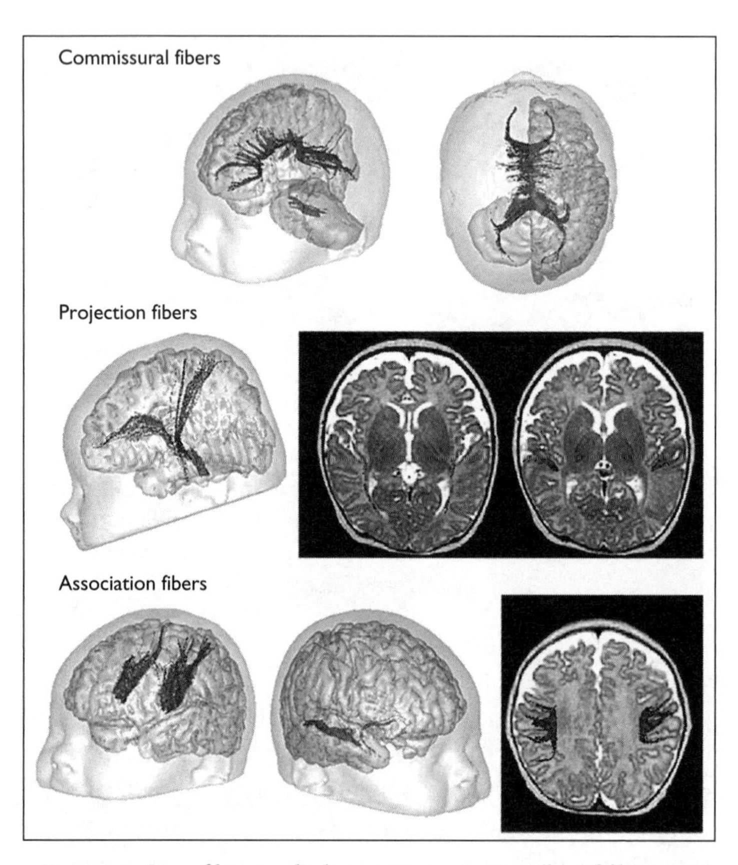

Figure 21: *Long fibers in the human neonate.* Bundles of fibers are shown by diffusion MRI. From J. Dubois et al., "Assessment of the Early Organization and Maturation of Infants' Cerebral White Matter Fiber Bundles: A Feasibility Study Using Quantitative Diffusion Tensor Imaging and Tractography," *NeuroImage* 30 (2006): 1121–1132; reprinted with permission from Elsevier.

evant context, a psychological structure, a subassembly of knowledge and hypotheses that the communicators have of the world. This can be information about their immediate physical environment, scientific hypotheses, religious beliefs, political opinions, cultural prejudice, and even suppositions about the mental state of the listener. This shared knowledge does not necessarily correspond to individual knowledge, as Grice pointed out. The theory of inferential communication is based on the intention of a speaker that his utterance produce a certain effect on his audience through the listener's recognition of that intention. The listener infers the intention of the speaker. There is therefore a need for cooperation between the participants for them to recognize a single common aim or a number of common aims. The cooperation is by conversation. In his 1967 lectures on William James, Grice went even further when he stated that the act of communication raises expectations that would be exploited in a defined cognitive environment. The inferential model of communication contrasts with an empiricist input-output model. It is in a hierarchical framework of contextualized prerepresentations of defined intentions, in which participants communicate in a projective style. Incidentally, the inferential model provides an explanation for the Chomskian paradox of the poverty of the stimulus and the richness of innate knowledge. As such, it raises an important question. How could a spoken word of modest intrinsic meaning produce an effect that mobilizes considerable long-term memory? In their seminal work, *Relevance,* Sperber and Wilson dealt with the efficiency of information processing, the minimum cost that ensures the maximum improvement of an individual's knowledge of the world. They argued that relevance was a measure of the multiplication effect created by the combination of new information with old information. The greater the multiplication effect, the greater the relevance. A process of ostension attracts the attention of the communicators to get together in reciprocal communication with a maximum of efficiency in a common framework of intention.

MIRROR NEURONS AND RECIPROCITY OF COMMUNICATION OF INTENTIONS

As we discussed in chapter 1, the systematic exploration of the physiological properties of single neurons in various motor areas of the monkey's cerebral cortex by Rizzolatti and his colleagues in 1996 led them to discover peculiar so-called mirror neurons. They are of interest in the context of communication between individuals, which we are discussing here. Their findings concern the premotor area of the frontal cortex. The neurons are active when the monkey reaches out its hand to seize food and bring it to its mouth. This is a complex motor task involving somatosensory and visual stimuli. The important observation is that these motor neurons are also activated by visual stimuli in the absence of motor activity—for example, when the monkey is at rest and the experimenter makes a grasping movement. There is a correspondence between the observed action and the actual motor act. Hence the name mirror neurons. This mirror response is very specific. For some neurons, grasping the food directly with the fingers elicits a response, whereas using a metal tool does not. In others, a response may follow rotation of the hand around a grape in one direction, for instance, but not in the other. There is thus a close relationship between an observed and an executed action for a given neuron.

Rizzolatti and his colleagues noted that the area in which they found their mirror neurons in the monkey is homologous to cortical area 45 in humans, part of the so-called Broca's area, that part of the cortex responsible for language, as we shall soon discuss in detail, in which there are also representations of the hand, along with those of muscles controlling the mouth and pronunciation of the spoken word. They proposed a relationship between mirror neurons and linguistic communication that is in agreement with Alvin Liberman's theory that perception of words involves visual perception of lip and face movements. Using human cerebral imaging, they confirmed and extended their electrophysiological observations in monkeys. Similar areas are activated when a motor task is executed or observed, that is to say, area 45 and temporal area 21. So there is reciprocity in recognizing motor acts in others, resonance between partners engaged in a mutual dialogue, and a capacity for imitation and therefore communication of intention. Nevertheless, mirror neurons represent only a small fraction of the neuronal systems involved in imitation and communication of intention, which, as we shall see, mobilize not only processes of recognition of others but the sharing of rewards.

THE NEURONAL BASIS OF THEORY OF MIND AND ATTRIBUTION

Humans are social animals in whom interaction between individuals is different from that in other species. Earlier I analyzed the conditions necessary for a tit-for-tat theory of cooperation as proposed by Axelrod in 1984. A cooperative strategy of a kind that would resolve the prisoner's dilemma game requires several conditions. These include mutual recognition by the partners, memory of earlier meetings, reward or punishment for cooperation or not, and a common meeting point. Unlike other species, humans also possess rationality and sociability, what one might call social intelligence. According to Chris Frith and Uta Frith in 1999, this further includes recognition of one's place in society, acquisition of knowledge from others, and teaching new skills and knowledge to others—in other words, pedagogy. Humans have a unique capacity to understand and manipulate the mental state of others and so modify their behavior, and they possess the awareness that others have knowledge, beliefs, and desires similar to or different from their own, by which they organize their behavior. As we saw earlier, these features can be grouped as theory of mind, or attribution, or what we might call intentional stance. This calculation of others' intentions, well known in political circles from Machiavelli to François Mitterrand, can now be analyzed in children. The Sally–Anne scenario of Simon Baron-Cohen and his colleagues illustrates the concept of false belief. Sally puts a marble in a basket and then goes out. Anne takes the marble out of the basket and puts it in a box. The subject children are asked where Sally should look for the marble. Children must understand that Sally will probably look in the basket because she has a false belief, a fact that the children must detect.

Neuronal correlates of theory of mind are numerous. Patients with certain prefrontal lesions, notably of the orbital and medial areas, have serious problems in social behavior. Furthermore, cerebral imaging reveals medial prefrontal, parieto-temporal, and inferolateral frontal cortical activation during tasks involving theory of mind. Recordings in primates demonstrate basic systems underlying theory of mind. There are distinct responses in the superior temporal cortex to hand movements and faces, but not to inanimate objects, and responses to the direction of others' gaze to guide one's own gaze. There are mirror neurons for motivated performance in the inferior prefrontal cortex, and no distinction between self and others. Responses distinguishing self from others are found in the superior temporal cortex; these responses differentiate between sounds or images coming from others but not from oneself, according to Perrett. There can also be activation of medial prefrontal and anterior cingulate neurons that anticipates the production of self-initiated movements, and thus suggests an explicit representation of self.

THE CONSCIOUS NEURONAL WORKSPACE: A NEURONAL MODEL OF SOCIAL NORMALIZATION

Our model (Figure 22) extends the neuronal hypothesis of the conscious neuronal workspace and distinguishes systems that operate *unconsciously* from the *conscious* workspace, where multimodal representations are formed on the basis of long-axon neurons, the cell bodies of which are mainly (but not exclusively) in layers II and III of the cortex. Such neurons are frequent in the prefrontal, parietal, and cingulate cortex, the very regions activated by conscious tasks. The workspace model is completed by a schema of reward sharing that enables social normalization of relations between signifier and signified. Saussure likened the signifier to, for example, a sound, and the signified to a conveyed thought. This could be through the intermediary of conversation, in which people speak in the same language in the same intentional context and with common, or different, representations. There could be stabilization of prerepresentations linking signifier and signified through propositions having a common shared meaning, as well as destabilization of nonpertinent prerepresentations.

Syntax and the logic of the proposition are progressively

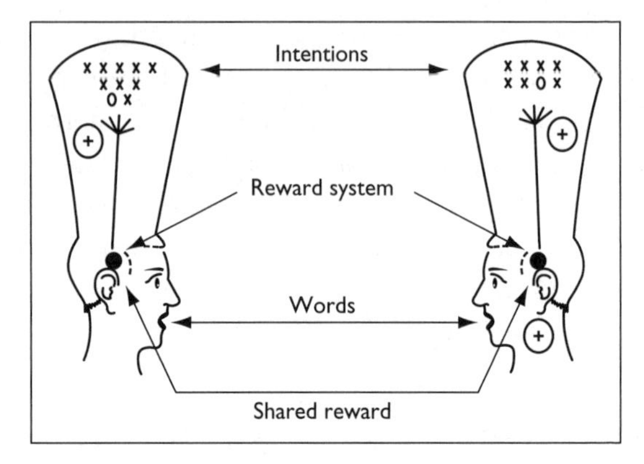

Figure 22: *Model of language learning by shared reward.* Adapted from J.-P. Changeux, *The Physiology of Truth* (Paris: Odile Jacob, 2002); reprinted with permission.

established in the course of social interaction. Many groups, including Helen Neville's, have used cerebral imaging and recording of evoked potentials to demonstrate differences between semantic and grammatical processing, and between nouns and verbs, in terms of the cortical areas used, although the left prefrontal cortex seems to prevail for most functions. As we saw earlier, there can be significant changes in event-related potentials during tests of semantic or musical congruence, depending on whether a phrase has a relevant meaning ("the pizza was too hot to cry"), or whether a melody is tonal or atonal or contains rhythmic delays that break the melodic contour (Mireille Besson). In all cases of incongruence the amplitude of the P600 wave increases. Does that indicate absence of reward or the perception of discord or even a punishment? Whatever the case, there exist neuronal correlates for the difference between sense and nonsense.

Olivier Houdé and his colleagues in 2000 compared cerebral images of subjects performing a perceptual task without making any particular effort and making frequent errors, and then performing the same task while making an effort to avoid errors and overcome perceptual bias that is due to learned logical reasoning. The images in the two cases differed spectacularly: the observers found activation of the posterior (occipital) cortex in the "effortless" task, but of the anterior cortex, including the left prefrontal and Broca's areas, with the more difficult, logical effort. The neural bases of logic, as well as epigenetic rules to organize it, acquired through learning, as we discussed earlier, are henceforth accessible to objective analysis by the neuroscientist.

The Neural Basis of Language

Saussure was a pioneer in formulating concepts of language, which can be seen both as the faculty of being able to express oneself using words and as the technical product of that faculty, which permits its exercise by individuals as a social vehicle for reason and emotion. Saussure emphasized the importance of words produced by a speaker and described a formal circuit based on them. Conscious facts or concepts trigger in the brain of speaker A an acoustic image (a psychic phenomenon), which is transmitted to the organs of phonation (a physiological phenomenon), which produces sound waves for propagation from A's mouth to listener B's ear. The circuit is completed by the concepts produced in B's brain. Language is form, not substance, but exists thanks only to some material support. For Saussure a linguistic sign unites not an object and a name but a concept and an acoustic image, a signified and a signifier. The link between them is, with few exceptions, arbitrary. Owing to their acoustic nature, units of language propagate along a chain of speech, each individual term being related syntagmatically to the preceding and subsequent ones. Perpendicular to this axis are a series of paradigmatic axes, along which are placed units, or paradigms, that can replace each other by commutation. He saw language as a formal system of signs, a system of relations comparable to the rules of chess, by which specific functions are allotted to individual pieces, yet with a neural basis.

The behaviorist theory of language was derived from John Watson's theories in his *Behaviorism* (1925), in which language was described as the sum of the "verbal habits" of an individual, and thinking as "subvocal language" exercised "behind the closed doors of the lips." Language and thought were reduced to behavior like any other behavior, but with "word substitutes" for every object in the environment, and "equivalence for reaction between objects and words" creating "economy of time" and the ability to cooperate with other groups. The idea was taken up by Leonard Bloomfield, who, in *Language* (1933), defined a "substitute linguistic reaction." In *Verbal Behavior* (1957) B. F. Skinner was more radical, reducing language to objective behavior, stimulus and response in communication. Language could be analyzed functionally rather than formally. Besides stimulus and response, Skinner took into account the action of the environment on an organism *after* a response has occurred, that is, the possibility of reinforcement. There could be inferences that might be realized. For Skinner the environment played a role in the selection of responses by the organism, and this led to his Darwinian concept of behavior and language acquisition by an individual.

Roman Jakobson of the Linguistic Circle of Prague and André Martinet in France completed the structural and functional description of Saussure's system of rules. A sender sends a coded message within a context to a receiver with three functions: expressive (centered on the sender), conative (oriented toward the receiver), and referential (concerning the contents of the message). A major contribution of the Prague School was the distinction between phonetics, the analysis of sounds, and phonology, the analysis of acoustic images. So there is a double aspect to language: first, a succession of concepts represented by signs, or monemes; second, perceptible phonic forms, or phonemes, all of which contribute to the best functional outcome.

A linguistic revolution took place when Chomsky published his works *Syntactic Structures* and *Aspects of the Theory of Syntax.* He emphasized the importance of accounting for the human ability to produce, engender, and understand infinite numbers of linguistic statements. This ability is inborn and determined biologically. Such generative power is "an innate set of linguistic principles shared by all humans." The environment has no intrinsic structure: there is only internal law and order, revealed by the linguistic experience. There is no general theory of learning, but a passage from an initial state, S_O, genetically determined in successive stages, to a stationary state, S_S, through experience. For Chomsky, language is an autonomous formal system that cannot be explained by its function. It is analogous to a series of instructions from a programmer to a computer, whatever the neural substrate might be, a dualist tendency. He described a generative grammar aimed at producing acceptable phrases: its creativity is bound by formal rules, consisting of three components: syntactic, semantic, and phonological. The syntactic component is subdivided into a basic component, which produces a deep structure, and a transformational component, for passing from the deep structure to a surface structure. The basic component includes a lexicon, and the rules of lexical insertion and the selection of the lexicon introduce semantics into syntax, whereas Chomsky's original proposal was that, on the contrary, semantics are deduced from syntax.

Piaget was more a psychologist and epistemologist than a linguist, but his theories had considerable repercussions owing to his interest in children and learning. For him language was ideal for revealing thoughts, and linguistic statements "translated" the mechanisms of intelligence. A child's first language is egocentric, emotive, and expressive, composed of monologues or verbal commentaries accompanying actions and gestures. From the age of five to seven a more socialized language appears, with referential and communicative functions, taking into account context and intentions, much like that of adults. The child progresses from egocentricity to socialization. Piaget's theory of knowledge, which was also a theory of learning, was based on the idea that "living organisms are active systems of response and reorganization" that "assimilate the characteristics of their milieu and in return adapt to it by reorganization." There is a transfer of order and even structure from the environment to the organism by a process of phenocopy, which has no real biological basis. This interactionism is matched by constructivism, according to which knowledge is gained by an array of choices and actions on the environment in the form of successive states of equilibrium. What is acquired from any given state is integrated in higher ones. The cognitive development of the child includes successive states of emergence of the first forms of abstraction (from birth to eighteen months), of representation and conceptualization (two to four years), concrete operations (four to five years), and "hypothetico-deductive" propositional logic (from eleven or twelve years onward). For Piaget language was only one of various manifestations of symbolic function. It is based on representations, first mental images, then symbols, and finally linguistic signs. Intelligence in action progresses to mental operations.

Vygotsky, Pavlov, and Luria stand out for the importance they attached to the role of language in social communication. According to Vygotsky in 1934, language began as a means of communication between adults, and between adults and children. It is external to the child in both form and function, but it is later interiorized to become thought. For him the child's egocentric language is, contrary to Piaget's theory, a form of social language.

Several biologists, including Niels Jerne, Edelman, and me, have tackled the question of Darwinian mechanisms in the acquisition of language, mechanisms based on the progressive development of cerebral connectivity and the selection of preexisting variations in this connectivity. Such theories are particularly applicable to learning a language, as Massimo Piattelli-Palmarini has pointed out. Despite their ingenuity and diversity, most language theories remain to be validated experimentally. The demonstration of coherence between a linguistic fact and a given theory is not enough to substantiate the theory as long as there is no evidence for causality. Causality implies a relationship between structure and function. We need to study the neural basis of language, particularly in terms of specialized areas of the brain.

LANGUAGE SPECIALIZATION IN THE HUMAN BRAIN

Sound communication exists in many animal species. In monkeys it takes the form of exchanges of vocalizations, and some scientists talk of "conversations" between chimpanzees. Of course, we have lost forever firsthand witnesses of such communication among our direct human ancestors, but we do possess anatomical data such as the imprints of the meninges and their vessels on the inner surface of skulls, as well as cultural evidence of habitations, fires, tools, sculptures, and paintings. From *Australopithecus* to *Homo sapiens sapiens,* cerebral volume has jumped from around 475 to some 1,400 cubic centimeters. But what is more important for our consideration is the reorganization of the posterior parietal cortex and the relative growth of the frontal cortex, already noted in *Homo erectus* by Ralph Holloway. In *Homo erectus* the asymmetry between the two hemispheres, described in *Homo habilis* by Phillip Tobias, was even more pronounced. For Holloway this reorganization of the brain indicated the development of visual and auditory communication and multimodal processing of information as a basis for language. This related to the development of socialization and its complex network of communication, visuospatial integration, and memory. It was necessary for the manufacture of tools and, later, of writing.

The volume of the human brain grows 4.3 times postnatally, compared with 1.6 times in the chimpanzee. Although morphogenesis of cerebral architecture takes a few months, synaptogenesis, the production of new synapses and the elimination of unwanted ones, is prolonged for years, into puberty. During this time the sociocultural and educational environments leave their traces in the developing neural networks. Postnatal synaptic plasticity enables the transmission of the cultural heritage of one generation to another, and its evolution.

Language serves first and foremost as a system of verbal communication within human populations. To recall famous metaphors of the nineteenth century, it was Spencer's "apparatus for conveying thought" within a "social organism," as described by Saint-Simon and Comte, through the communication of signals, knowledge, and rules between different partners of a social group. Language is a symbolic system of variable and arbitrary signifiers, and it serves to transmit the signified through basic syntactic rules, for the most part common throughout, and specific to, humankind. So we should distinguish between the characteristic human genetic envelope, which makes all the difference between humans and chimpanzees or macaques, and epigenetic variation that is related to the exceptionally long postnatal development of humans.

Darwin noted that the evolution of languages could be compared to that of living creatures and be expressed in the form of a phylogenetic tree. The comparative analysis of the great linguistic families by Joseph Greenberg and Merritt Ruhlen revealed diversification by geographical isolation that could also be expressed in the form of a phylogenetic tree of languages. Even if the circular diagrams proposed by Claude Hagège and his colleagues demonstrated fusion, death, and rebirth within languages, the comparison of linguistic and genetic data by Luigi Cavalli-Sforza and his collaborators tended toward a "demic," or migratory, evolution of languages. They observed a parallelism between the evolution of neutral genetic markers and the evolution of languages, customs, religions, and moral rules, which suggests that humans migrated with their cultural baggage rather than spreading their language "epidemically" from individual to individual. There was a rapid and epidemic spread of knowledge, whereas learning of language was slow and difficult to reverse.

In the monkey we find cortical areas homologous to human language sites, such as Broca's area, but they do not have the same function. Francisco Aboitiz and Ricardo García in 1997 described a large neurocognitive network of language-related circuits that includes interconnections among temporal, parietal, and frontal (especially prefrontal) cortex, which enables the development of multimodal associations. Wernicke's area, in the posterior part of the temporal cortex and the adjacent parietal cortex, could be the site for convergence of conceptual associations and their phonological correlates. These phonological representations could then correspond to more inferior temporal areas and thence to Broca's area, forming a circuit for working memory and the handling and learning of complex vocalization. Insofar as this system can access the prefrontal cortex (and so the conscious neuronal workspace), it becomes able to produce more and more complex structures, the first forms of syntax. In parallel, there is differentiation of the hemispheres, the right dealing with iconic or "pictogenic" analysis, and the left with verbal and semantic computational analysis.

NEUROPSYCHOLOGY OF LANGUAGE

In the early nineteenth century Franz Gall proposed the existence of cortical areas devoted to language. Marc Dax and Jean-Baptiste Bouillaud had clinical evidence for loss of speech associated with lesions of the frontal lobes of the brain. The first real neuropsychological analysis of language, however, began with Broca in 1861 and Ludwig Lichtheim in 1885. Broca convinced the scientific world of the special role of the left hemisphere in language production. He described his patient, Leborgne, and a few others as suffering from aphasia, retaining their comprehension of language and reasoning, but having great difficulty actually speaking. In such cases there were lesions of the posterior part of the left frontal cortex. Later analysis showed dissociation between verbal production of language

(Broca's area), and comprehension of words. The problem posed by the development of a tree of the signified, with parallel and hierarchical organizations, is to find the association between this neural organization of semantics and the repertory of auditory signifiers (and visual ones in the case of reading). In 1874 Carl Wernicke described an area in the left parietotemporal cortex, and a rather curious disorder in which the pronunciation of words was preserved but there was difficulty choosing words, and a jargon totally devoid of syntax. He proposed an anterior motor center, lesions of which cause Broca's aphasia, whereas his posterior sensory auditory center was associated with what we now call Wernicke's aphasia. In 1885 Ludwig Lichtheim, the best known of what we might now call "modelers," added a "concept center," which receives messages from the verbal auditory center and sends them to the motor center. Lesions cause mind blindness (Heinrich Lissauer, 1890) or agnosia (Sigmund Freud, 1891). He proposed that an interruption of fiber tracts linking the two centers leads to transcortical aphasia, either sensory or motor. Jean-Martin Charcot in 1885 and Joseph Grasset in 1896 adopted these models, emphasizing that the centers should be in the gray matter, and modern connectionist cognitivists, such as John Morton in 1980, have accepted them.

The idealist reactions of Bergson, and more seriously the theories of Hughlings Jackson and Henry Head, opposed these localizationist concepts. As we noted earlier, Hughlings Jackson, perhaps inspired by Spencer's ideas of the evolution of the brain, distinguished successive levels of organization in the brain, from simple to complex, from rigid to labile, from automatic to voluntary; each higher level controls more complex aspects of behavior across the lower levels. So lesions of higher levels produce dissolution, the opposite of evolution. He saw dissolution as revealing several forms of language, such as emotional, inferior, and superior. Higher centers are the most autonomous and have the ability for "debate" among themselves. They can be the seat of new arrangements, more or less ephemeral and fragile, and thus provide a basis for internal evolution. For Head in 1921 there was no language center but a "preferential integration focus," lesions of which disorganize an ordered sequence of physiological processes.

Neuropsychological methods of analyzing the consequences of lesions were underpinned in the 1950s by Penfield and his colleagues using cartography of language areas by electrical stimulation of epileptic patients before resection of epileptogenic foci in the cortex. In the resting patient, most cortical areas were silent, except for primary sensory or motor areas. If a patient was speaking, mild electrical stimulation of Broca's or Wernicke's area, or certain nearby areas, caused a transitory aphasic state. In 1983 George Ojemann used similar methods and showed that stimulation at certain precise sites led to an inability to name objects, notably in one language but not the other in a bilingual patient. Different sites were active in orofacial mimicry, in short-term memory, or in syntax. Interestingly, there was much individual and sexual variation in cortical localization of language. Linguists such as Jakobson and Jean Gagnepain and Olivier Sabouraud have attempted to interpret aphasia on the basis of linguistics. For the latter two, as for Karl Lashley, human behavior is based on a serial temporal order related to the planning ability of the frontal cortex, which intervened in the mediation between two evolving systems: the nervous system and culture. Aphasia affects especially this power of mediation, so Broca's aphasia alters the generative capacity for syntactic encoding. This causes, at the semiological level, agrammatism, which includes disturbances of contiguity but preservation of taxonomy, and, at the phonological level, errors in choice and use of phonemes. Wernicke's aphasia affects paradigmatic decoding, which results in problems of word choice and the use of a jargon full of verbal neoformations unknown to any vocabulary.

Neuropsychological analysis demonstrates clearly the difference between comprehension of words and their production. First, comprehension can be disturbed at the level of primary acoustic processing (as occurs when there are lesions of the temporal lobes) or the processing of phonemes in the left hemisphere. Then there can be defects in understanding the meaning of words—for example, colors, body parts, concrete or abstract concepts, actions, or proper names (sometimes being limited to historic personalities). All seems as if the Saussurian signifier were selectively dissociated from the signified. Word production can be selectively disturbed by lesions, which cause errors in selection and ordering of phonemes, or kinetic faults in rhythm or accentuation. In his 1995 book, *Language and Language Disorders,* Sabouraud reexamined classic aphasia, concluding that it could not simply be interpreted in terms of Saussure's linguistic signs. Wernicke's aphasia produces a jargon consisting of series of syllables whose tone, delivery, and prosody are preserved, but which contains incongruous words, absurd phonic sequences, and neologisms, all without the speaker's being aware that anything is amiss. The abundant verbal flow is without syntax, phrases have no structure or proper endings, and there is no "solidarity" between words. Such problems with language production are accompanied by similar troubles of understanding. Patients are confused about meaning and respond poorly or not at all to simple orders. For Sabouraud linguistic capacity follows two axes, the taxonomic (lexical) and generative (textual). So Wernicke's aphasia would be a selective change in taxonomic capacity, which produces a deficit in constructing words from their parts, that is to say, assembling the signified. Broca's aphasia is different: the consequence is mutism, or at best reduced verbal usage accompanied by rare and brief interjections, often stereotypical. Patients cannot improvise the simplest of phrases. Their capacity to generate the infinity of expressions that make up language is destroyed.

Aphasia can be differentiated from other disturbances of language. Mental confusions, such as the delirium of a feverish child, or of a drunkard or the senile, are characterized by a defect of memory and selective attention, and a denial of reality. Frontal lobe lesions can lead to incoherent or stereotyped speech, lack of initiative, inappropriate use of objects, and distractibility. In terms of speech, there is no loss of syntax, but rather a failure to produce sense.

Epigenesis of the Sign

ALLIANCE OF SIGNIFIER AND SIGNIFIED

The science of signs (something that means something to someone), or semiotics, really began with Charles Sanders Peirce (1839–1914). He was an autodidact, chemist, engineer, son of a mathematician, and friend of William James. He was not recognized in his lifetime, however, by the philosophical and linguistic communities. He classified signs as a function of their effect on an observer. These were effective (the feeling or proof that we understand), energetic (the physical, muscular effort involved), or logical (the mental effect involving a change in habit). For Peirce human experience is nothing if isolated. The universe of signs is a community, a democracy where *interpretants* are not personal property but common property transmitted by language and guarantees of reality and truth: "The opinion which is fated to be ultimately agreed to by all who investigate, is what we mean by the truth, and the object represented in this opinion is the real." Logic has its roots in the social, and man is the most perfect sign.

A little later Saussure attempted to establish an exact science of signs, like a systematic social science uniting the sociology of Emile Durkheim (1858–1917) and the mentalist psychology of the time. He distinguished three levels: the social community, the individual, and the language. He saw language as a system of signs centered on words, whereby the sign unites not an object and a name but a concept and a sound image. It is a two-sided physical entity uniting two mental images. On the one hand is the *signified,* that is, the concept that is the representation of the knowledge that a subject has of an object and which contains a subjective content. On the other hand is the *signifier,* that is, the acoustic image that is representative of the sound sequence.

SYNAPTOGENESIS AND EFFECTS OF LEARNING DURING DEVELOPMENT

Learning a language is essentially postnatal, even if certain memory tracks can be laid down before birth. In humans this postnatal development is particularly important and long. We might recall here that a human baby's cranial volume increases more than fourfold after birth, whereas that volume increases less than twofold in the chimpanzee, though their gestation periods are similar (270 days in humans, 224 in the chimpanzee). The human cranial volume reaches 70 percent of the adult value after three years, whereas it reaches that value after one year in the chimpanzee. Nevertheless, postnatal brain growth is not restricted to humans: the brain grows almost six times in the rat.

In 1993 Jean-Pierre Bourgeois and Pasko Rakic undertook a systematic quantitative electron microscope study of the macaque to evaluate the evolution of synapses in the visual cortex. They described five phases. The first phase involves precortical synaptogenesis in the primitive layers of the so-called pallium, where the cortex eventually develops. This phase includes the marginal zone and subplate and occurs some sixty days after conception. Then comes an early, truly cortical phase of synaptogenesis in the cortical plate, coinciding with a peak of neurogenesis seventy to one hundred days postconception. Then begins a rapid cortical phase that lasts until two months after birth. This is the fastest period of synaptogenesis, as 40,000 synapses per second develop, paralleled by increased production of dendritic spines, which are minute thornlike protuberances from the dendrites; these receive the bulk of excitatory inputs to cortical neurons. There is a plateau from infancy to puberty, during which there is a maximum density of 600 to 900 million synapses per cubic millimeter. Then comes a phase of decline up to adulthood, in which there is loss of synapses on the spines, and a final, rapid decline in senescence. Similar phases have been described in humans, but with differences between cortical areas. The evolution of the prefrontal cortex is longer, at ten years, than primary visual, at two to three years, and begins later. Furthermore, in the macaque synaptogenesis differs between cortical layers. Spine development is continuous in layer III, but transitory in layer IV. Certain phases, notably the third, fourth, and fifth, are influenced more by experience. The length of the third phase increases from 14 days in the rat, to 30 in the cat, 136 in the macaque, and 400 in humans. This epigenetic heterochrony, the progressive lengthening of phase 3, allows for an increase in the number of epigenetic combinations by lengthening the postnatal period of plasticity. It ensures a prolonged sociocultural interaction suitable for the acquisition of language with only a modest cost in genes.

What is the role of neurotransmitters in this development of the brain? Does it happen in the absence of transmitters? In 2000 Matthijs Verhage, Thomas Südhof, and their colleagues decided to delete the gene Munc18–1/ns 1, which codes for a protein necessary for synaptic activity. After this deletion, mice could no longer release transmitters, although the structure of their brains and neuromuscular junctions appeared normal. There were, however, signs of massive neurodegeneration, especially in regions that developed early, such as the brainstem, although in structures that developed later, such as the cortex, there was no obvious difference between wild mice and the mutants. So it seems that the activity produced by transmitter release is necessary not for the formation of neural structures, but for their maintenance or stabilization.

My colleagues and I suggested a theory of epigenesis by selective stabilization of synapses in 1973, and it seems fitting to reexamine it nearly forty years later (Figures 23, 24). Bearing in mind subsequent advances, we might retain five principles. First, a genetic envelope determines the primary traits of the anatomical organization of the brain of a given species, which, although not needing transmitter release for their formation, can be changed by genetic mutations. They include morphogenesis of the neural tube, from which the brain develops in the embryo, and the brain itself, migration and differentiation of nerve and glial cells, development of synaptic connectivity, spontaneous activation of neuronal networks, and the evolution of networks by ongoing activity. Second,

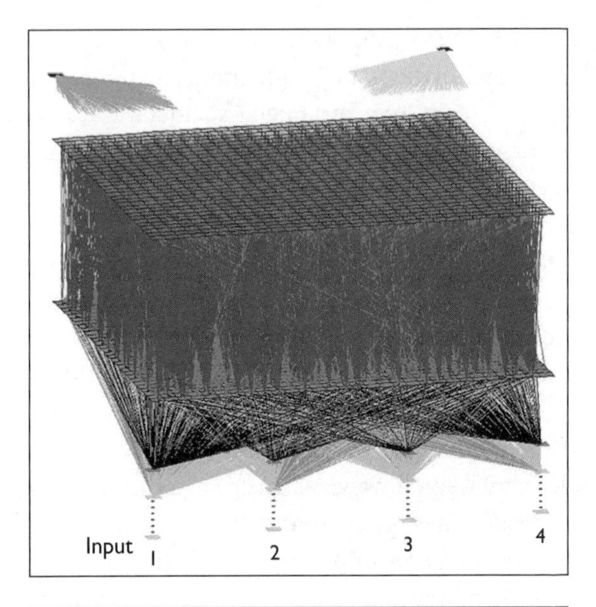
Input
1
2
3
4

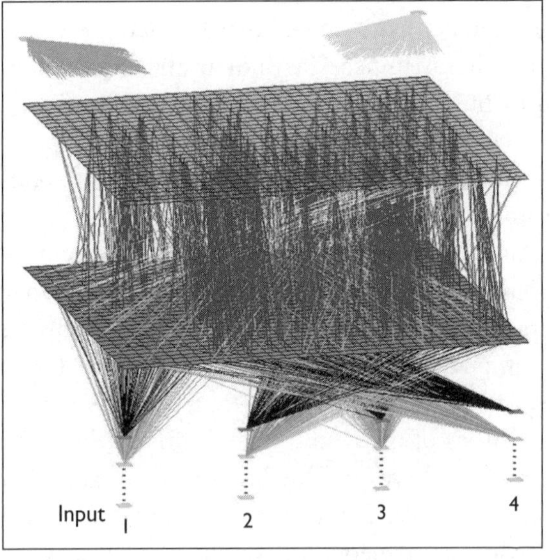
Input
1
2
3
4

there is phenotypic variability of adult neural organization even in isogenic organisms, and it increases as one progresses from invertebrates to man. Third, synaptic connectivity progresses by successive waves of exuberance and regression, and there is a critical period for each wave when connectivity is at a maximum. Fourth, nervous activity within a network, initially spontaneous and then evoked by interaction with the environment, regulates stabilization or elimination of synapses during the critical period. Finally, in the adult, we find that the phenomena of neurogenesis, synaptogenesis, and stabilization persist, but to a limited extent.

Synaptic evolution progresses reversibly from labile to stable, and both of these can regress to degeneration, which is one-way only. The neuronal program, which includes maximum connectivity, the principal stages of network development, the modalities for stabilization of labile synapses (including retrograde signaling), and a neuron's capacity for integration, is determined genetically. An evolution equation describes the evolution of connectivity through the total message of afferent activity to the postsynaptic cell during a given time period. The development of neuronal individualization, or singularity, is controlled by the activity of a forming network that orders the selective stabilization of a particular distribution of synaptic contacts from among those present at the time of maximum diversity. A new proposal is that a terminal amplification

Figure 23 (opposite): *Epigenetic reward-based selective stabilization of synapses.* Computer simulation showing, *top,* numerous diffuse connections before learning and, *bottom,* fewer organized, coherent connections after learning. From T. Gisiger et al., "Acquisition and Performance of Delayed-Response Tasks: A Neural Network Model," *Cerebral Cortex* 15 (2005): 489–506; reprinted with permission from the Oxford University Press.

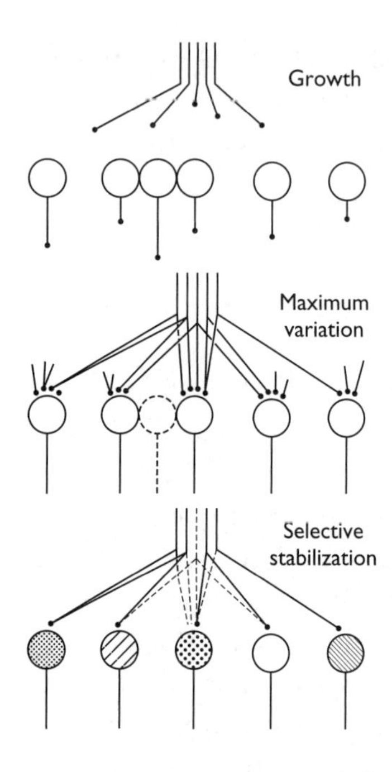

Figure 24: *Selective synaptic stabilization,* shown during growth, at maximum innervation, and after activity-related stabilization. From J.-P. Changeux, *The Physiology of Truth* (Paris: Odile Jacob, 2002); reprinted with permission.

of selected pathways can take place if the stabilizing activity is maintained. Such a model has two major applications. First is a temporal distribution of nerve impulses in the form of a stable track that can be described in terms of synaptic geometry. Second is a phenomenon of variability, according to which the same input message can select different connectional organizations but lead to the same input-output relationship—that is to say, to the same behavior by the organism, in spite of the totally deterministic character of the model.

Two alternative models to that of epigenesis exist. In the "innatist" model, Chomsky considered that mental organs are determined genetically and are suited to a given species. Intrinsic psychological structure is rich and diverse, and the stimulus poor. This model, however, does not account for the critical effects of experience and the creativity of language, unless one adds to it mechanisms of selection. As to the empiricist model, for associationists from Aristotle to Hilary Putnam, everything in the mind begins in the senses. Practical usage reinforces connections and nonusage weakens them. According to Steven Quartz and Terrence Sejnowski, the environment is structured and guides development, an idea that reflects a sort of neuronal Lamarckism. But this model does not take into account the importance of a genetic envelope in language acquisition.

The general model of synaptic selection has been improved by plausible molecular mechanisms involving either growth factors or inhibitory systems. Jean-Luc Gouzé and his colleagues proposed competition for a retrograde signal, a trophic factor produced in limited amounts by the postsynaptic neuron and taken up actively by competing nerve terminals. In a similar vein, Lamberto Maffei demonstrated that nerve growth factor, administered during a critical period of experimental visual deprivation, reduced the effects of deprivation on the development of the visual cortex, in particular the disturbed formation of ocular dominance columns. Furthermore, cortical neuronal networks contain small inhibitory interneurons in addition to excitatory neurons. The common transmitter in the cells is gamma aminobutyric acid (GABA). Muscimol is a GABA agonist and also diminishes the effect of visual deprivation on the selectivity of visual cortical neurons. A mutation in glutamic acid decarboxylase, an enzyme in the synthetic pathway of GABA, leads to a decreased effect of visual deprivation in cortical development.

Detailed analysis by fMRI and transcranial magnetic stimulation of the cortex by externally applied magnetic fields during learning to read Braille by congenitally blind subjects has also demonstrated an increased somatosensory representation of the "reading" hand and increased activity in the primary and secondary visual cortex that cannot be receiving direct visual stimuli. So the visual cortex is being recruited for spatial tactile tasks in the absence of visual stimulation. Further, transcranial stimulation of the visual cortex interferes with reading Braille in such a way that the subject cannot say if the text has any meaning. The simplest model to explain these results is that corticocortical connections exist at birth between somatosensory and visual cortex, and between nonvisual thalamus and visual cortex. In the congenitally blind child there will be, as Braille is learned, selective facilitation of these pathways, which will benefit tactile perception of writing.

The Cerebral Imprint of Writing

Writing is an exclusively human invention; its evolution was made possible by the existence of both innate predispositions and epigenetic plasticity, characteristics of the developing nervous system. It is a recent invention, dating to some 3,500 years BCE, after *Homo sapiens sapiens* had existed for around 100,000 years.

The words *write* and *inscribe* are from the Anglo-Saxon *writan,* to inscribe on bark, and the Latin *scribere,* to trace characters. The Indo-European root is *sker,* to shear, cut, or incise. In addition, the word *graph* can be related to the Indo-European *gerebh,* to scratch or scrape. So to write means to incise or scrape, make a track (extracerebral) in a medium such as stone, bone, or pottery, which it will retain like a memory. From there we can get to Greek *gramma.* In Arabic *kitab* means a book, derived from the idea of an assembly of letters. Finally is the Nordic *runes,* derive from the Old English *rún,* a secret or whisper. Thus, we can see writing as a track in some stable material bringing together signs that have a meaning for those who possess the code.

WORDS AND WRITING

According to Saussure, "Language is a system of signs expressing ideas," and these signs are "associations ratified by common accord" and "realities of which the seat is in the brain . . . an individual needs to learn to understand its use." A person does this after long experience and imprints in his brain, rather like identical copies of a dictionary spread among many people. As we have already seen, the linguistic sign arbitrarily unites a concept and a sound image, the signified and the signifier. The semiotic triangle of Charles Ogden and Ivor Richards in 1923, which is similar to Peirce's ideas, linked the following:

> *Thought or reference,* the realm of memory where recollections of past experiences and contexts occur
> *Referent,* a perceived object that creates the impression stored in the thought area
> *Symbol,* the word that calls up the referent through the mental processes of the reference.

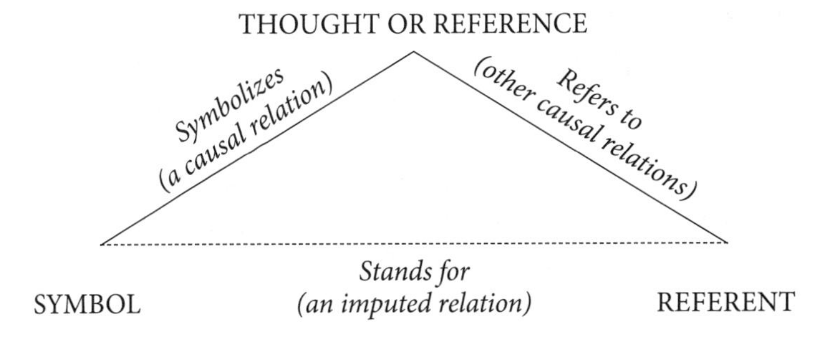

This triangle, however, does not account for two aspects of language. First, language is not made simply of names of objects: concepts are transformed and evolve. Second, how can concepts deposited in brains with variable connectivity code the same meaning from one individual to another? In his *Philosophical Investigations,* published posthumously in 1953, Ludwig Wittgenstein took up the question and concentrated on the criteria for the identity of two representations. He concluded that the significance of a word is its use in language. A child acquires language through language games with at least one partner, such as the mother, and at least one object in common. There is identification of the signified in cognitive games that involve divided attention and designation of a common object, then memorization and aural communication of the signified by the signifier, first pronounced by the mother and then repeated by the child. During these first cognitive games, and language games that follow, subjective validations are made that result in shared comprehension between partners.

According to Saussure, the only reason for writing to exist is to represent language. In his eyes, it enjoys undeserved importance; he spoke of the inconsequence of writing and the tyranny of letters. His point of view was not shared by Jack Goody, who in 1977 stated that writing has two principal functions: first, storing, marking, and recording information in order to communicate it over time and space; and second, ensuring passage from the aural domain to the visual, thus permitting a different way of examining things, of decontextualizing, by removing a word from the body of a sentence, from the flow of oral discourse, and therefore profoundly reorganizing oral communication. The introduction of pictographic writing extended the visibility of an act, whether judicial, political, religious, or indeed scientific. The graphic sign reaffirms the personal, the individual. With the advent of calligraphy it became a signature.

PREHISTORIC WALL PAINTINGS

Homo sapiens painted signs on the walls of caves more than 35,000 years ago. I use the word *sign* here in a general sense, as did Roman Jakobson (1896–1982), of its being related to a meaning, or a means of transmission between a sender and a receiver (information theory). There exist rules of transformation that relate a message to a given sign. Deciphering prehistoric signs remains, and will doubtless always remain, very hypothetical. André Leroi-Gourhan (1911–1986) noted that the first known prehistoric patterns consisted of regularly spaced incisions on bones or stones. Were they tales of hunting, exercises in decoration, or rhythmic incantations perhaps associated with dancing, similar to those found on the sacred churingas of Australia? Heads of animals began to appear about 30,000 years ago, associated with sexual symbols and negative outlines of hands. The animal art of the caves of Chauvet and Lascaux unites realistic figures of animals, such as bison, deer, and mammoths, with explicit abstract signs, such as dashes, lines, dots, and rectangles. Leroi-Gourhan recorded geometric patterns symbolizing the sexes. Between 8,000 and 9,000 years ago wall paintings disappeared in France, but abstract signs persisted in the form of painted stones at Maz d'Azil. Could this be a form of embryonic writing? The whole question of the interpretation of these prehistoric signs remains enigmatic. The negative images of hands have provoked much discussion, as some of them are mutilated. Could these be voluntary mutilations, accidental loss of digits, or a code based on fingers being bent or altered? Leroi-Gorhan suggested a secret code used by hunters. On the basis of rather scant evidence linking the occurrence of certain "mutilations" and the frequency of particular animal figures, he proposed, for example, that a complete open hand signified a horse; if four fingers were missing, it represented a bison. Thus were recorded the results of hunting on the cave walls. So the hands represent not sounds but actions. Could this be an early form of a relationship between a graphic signifier and the signified?

THE INVENTION OF WRITING

It is widely accepted that writing first appeared in Mesopotamia, but recent work has revealed earlier forms in India. There is evidence of *Homo erectus* near the Sea of Galilee a million and a half years ago; then *Homo sapiens* appeared in nomadic populations of hunter-gatherers. The first huts appeared around 15,000 BCE, and the first villages, 12,500 BCE in Jordan. The Natufians became sedentary, constructed food storerooms, and used mortar to construct houses. One can find here the earliest symbolic figures for women and bulls. In about 9000 BCE agriculture was born with the domestication of wheat and barley. Clay was being fired in Jericho. Skulls with plaster masks bear witness to territorial settlement of family groups. The next 2,000 years saw the domestication of goats, sheep, oxen, and pigs. Villages grew in size and harbored wall paintings, statues of animals, and fecundity symbols. Around 6000 BCE we find ceramics with simple geometrical patterns, such as those at Hassuna, then more figurative decoration using fish, birds, and human figures, such as those at Samarra, both in Iraq. Villages expanded and consisted of similar brick houses. Grain was stored collectively. Social organization was egalitarian, and there was solidarity. Accounting emerged: early calculi, tokens used for calculation, have been found at Qa'lat Jarmo in Kurdistan, perhaps the first signs of a human material code. Around 5000 BCE hierarchical urban societies were common. Intensive arable and livestock farming developed, and irrigation was needed. The economy became tentacular, according to Jean Bottéro. Large monumental edifices appeared among the houses. A mercantile upper class emerged. In Susa, in Iran, clay envelopes dating from 3000 BCE contained calculi, their surfaces were marked according to their number, and each was "authenticated" by seals. They could be details of coded contracts between parties. These clay envelopes became progressively empty and flat, the surface notes being enough: thus the clay tablet.

PICTO-IDEOGRAPHIC WRITING

Dating from about 3000 BCE, the ruins of Uruk in Sumer bear witness to a spectacular development of urban life. It covered 250 hectares and contained as many as 50,000 inhabitants, its *White Temple,* eighty meters long, and a palatial complex for the king. Long-distance commercial exchange expanded. The archives contain thousands of clay tablets in a Sumerian language, composed of figurative and schematic pictograms, which illustrate the stylistic and symbolic nature of all writing. These early pictograms tended to imitate an object: a cow was a triangle with two horns, a sheep was a cross in a circle (representing its enclosure). Sumerian pictograms could represent objects or concepts: they were ideograms. As is the case in Chinese, a sign represents a syllable. There was isomorphism, but the significance could go well beyond the pure figuration. Of 1,500 pictograms identified, most refer to personal objects or state administration in terms of treaties and laws, so this early writing had obvious utilitarian and commercial aims. Few had religious or literary connotations.

CUNEIFORM WRITING

The cuneiform system evolved technically and functionally during the third millennium BCE. Drawing realistic pictograms in soft clay demanded skill and precision. Scribes discovered that triangular impressions could be made using a reed stylus with a wedge-shaped end. The Latin *cuneus* means a wedge. Curves disappeared and signs became simple, schematic, and abstract. There were successive waves of immigration from Mesopotamia by the Akkadians, seminomadic tribes of Semitic origin coming from Arabia who introduced another language, and writing evolved further. Sumerians and Akkadians lived side by side in urban communities, politically independent but without hostility. Sumerian was a monosyllabic language; Akkadian was polysyllabic and featured declension and conjugation. The Akkadians used Sumerian ideograms, and they gave certain cuneiform signs phonetic value, thereby producing phonograms. Akkadian writing in the second millennium BCE became extremely difficult, mixing archaic ideographic signs with phonological innovations.

So there arose in Mesopotamia an authentic form of writing, from the pictogram, isomorphic but with signified richness, to the ideogram (with its abstract and formal global significance), independent of sound. Sound came with the *syllabogram,* and the grapho-semantic neural pathway was completed by a grapho-phonic path. Functional multilingualism that resulted from cultural mixing contributed to the evolution of writing. In *Race and History* Lévi-Strauss emphasized the innovative character of the "coalition of cultures." The willy-nilly introduction of new partners created diversification, which was followed by selection. In contrast to linear and constantly progressing "sociological evolutionism," Vico's three ages, Condorcet's unlimited progress, or Comte's three states, Lévi-Strauss proposed an authentically Darwinian model of cultural evolution. But for the sake of completeness we should recall that certain specialists, such as Pierre Encrevé, challenged the evolution of writing from the figurative to the abstract symbolic. They suggested that the abstract signs in wall paintings or ancient Susian pottery were the real precursors to writing, which has always been abstract.

Evolution of the Sign

Other forms of writing have evolved since antiquity independent of cuneiform, sometimes with notable differences.

EGYPTIAN

Hieroglyphic writing is documented before 3000 BCE in the Nile Valley, a little later than Sumerian pictograms, and was both picto-ideographic and phonetic. It changed little except that in the second millennium BCE there were 700 signs, and in the Greco-Roman period 5,000. Each sign was a stylized figurative design that could signify what was represented (a pictogram of a bull's head for a bull) or what it symbolized (an ideogram of two eyes for life), or a phonogram for pronunciation. There was an alphabet of 24 signs. The large number of hieroglyphs can be judged from the fact that Jean-François Champollion in the early nineteenth century deciphered the Rosetta Stone from 1,419 hieroglyphs and 486 Greek words. Egyptian writing presented categories of signs similar to those of classical Sumero-Akkadian. It is likely that a long, but unknown, evolution preceded hieroglyphics. Why the language evolved little during several millennia is also puzzling. Was it the privilege of the small percentage of literate people, or did the priests bring their magic to bear? Whatever the case, cursive writing in ink on pottery, wood, leather, and, especially, papyrus appeared in parallel. It was hieratic, written from right to left and following the same rules as hieroglyphs, though with a different graphic notation, or demotic, a later and more popular form. Curiously, the graphic, or calligraphic, style of early Arabic manuscripts at first sight resembles early demotic writing, although they are radically different in concept: alphabetic versus hieroglyphic.

CHINESE

In China the evolution of writing was similar to, but later than, the development of the first Sumerian pictograms. The first signs of authentic writing appeared engraved on oracle bones of the Shang Dynasty in Henan Province, north of the Yellow River, dating perhaps from the fourteenth century BCE. These bore witness to osteomancy, by which a diviner answered questions about the future from the king or generals or aristocrats. He applied an ember to the inner surface of the bone until a crack appeared, often in the form of 卜, the orientation of which was the diviner's answer, and it was noted with written signs. Thousands of such documents are known, and more than 1,000 signs have been identified. They are pictograms with schematic fish, horses, trees, bulls, mountains, and even ears. Later such signs appeared on bronzes, and they evolved between the twelfth and fourth centuries BCE from figurative pictograms to stylized ideograms. These ancient signs evolved into several basic styles. *Seal script* (小篆 Xiaozhuan) was harmonious and legible, within a rectangle. In the third century BCE Li Si established the form of some 3,000 such characters. *Clerical script* (隶书 Lishu) was more elegant and written with a bamboo or silk brush. *Regular script* (楷书Kaishu) and *semi-cursive script* (行書 Xingshu) were written with single or successive brushstrokes for esthetic pleasure. Eight fundamental strokes sufficed for 8,000 characters in the first century CE and 55,000 today, of which 3,000 are in common usage. The combination of strokes was neither totally arbitrary nor totally systematic. Originally pictograms represented objects, and symbols signified concrete or abstract ideas. Derived figures combined two characters to express a notion such as *light,* from the sun above a tree. Phonic characters were composed of an idea plus an indication of pronunciation; for example, *brilliant* was expressed as the sun plus a phonetic indicator. The fact that the same characters were used to transcribe the numerous spoken varieties of Chinese complicated matters.

KOREAN AND JAPANESE

Chinese writing spread to neighboring countries, Korea from the first century CE and Japan in the sixth century. As occurred when Sumerian and Akkadian combined, monosyllabic Chinese had to evolve to transcribe polysyllabic languages. In Korea an ideal simplification was decided by King Sejong, who in the fifteenth century systematized Chinese and removed its intrinsic meaning. He created Hangul, comprising twenty-eight vocal and consonant letters, so that previously illiterate Koreans, who did not have access to the existing Hanja language, could read and write.

Japan developed a mixed system of writing of extreme

complexity. Chinese kanji characters were retained, but syllabic katakana signs were added to transcribe words of foreign origin, and hiragana for words in common usage. There are about 3,000 kanji characters in common usage, and Latin letters are being progressively introduced into Japanese writing.

The Birth of the Alphabet

In the second millennium BCE the alphabet appeared around the Mediterranean, in Syria, Lebanon, Israel, and Jordan. Two major writing systems coexisted: cuneiform Sumero-Akkadian and *Protosinaic.* At Ugarit, near Byblos in Syria, cuneiform ideograms gave way to an alphabet of thirty cuneiform signs representing consonants, an acrophonic simplification of Akkadian: only the initial of the Akkadian syllable survived. At the site of Serabit el-Khadim, near the turquoise mines of the Sinai exploited by the pharaohs and in which Semitic workers were employed, graffiti has been found that used pictographic signs similar to Egyptian hieroglyphics. They do not seem to have had the same significance, and they appear to have been written in a different, Protosinaic language.

On the Mediterranean coast of Lebanon lived the Phoenicians, Semitic peoples related to the Aramaeans and the Hebrews. They founded independent cities and intensive commercial enterprises. In the tenth century BCE they invented the Phoenician alphabet of twenty-one consonants, perhaps derived from Protosinaic. Each pictographic letter was an acrophonic simplification. For example *aleph* ≮ would be derived from the bull's head pictogram, which after rotation became A. From this Phoenician alphabet sprang two main lines: the first was the Aramaean alphabet from the tenth to thirteenth centuries BCE, then "square script" Hebrew (first century BCE) and Arabic (fifth century CE); the second was the Greek (ninth century BCE), Etruscan (700 BCE) and Latin (500 BCE). Vowels appeared with the formation of the Greek alphabet. There again, the mixture of cultures and multilingualism caused writing to evolve. Indo-European languages were rich in vowels, whereas Semitic languages had mainly consonants. The Phoenician alphabet was enriched by vowels necessary for transcribing Greek.

This evolution of writing constitutes one of the best-documented examples of cultural evolution, presenting as it does the characteristic features of Darwinian evolution. Writing developed to store, make available, and transmit knowledge, accumulated by human society, which was too vast to be stored in individual memory. The graphics of language, in the West as well as the East, appear to have evolved from the figurative to the abstract, from the pictogram to the ideogram, and from the representation of sense to that of sound, from the phonogram to the alphabet. If certain writing systems evolved little over centuries, like Egyptian hieroglyphics and classical Chinese characters, others provided major advances, mainly by hybridization or cultural mixing. They came as Darwinian generators of diversity giving birth to Sumerian and Akkadian syllabograms, then to Egyptian and Semitic consonantic alphabets, then to Semitic and Indo-European vowels and the Korean alphabet. Existing systems of writing were used to transcribe new languages, and so they evolved. Constraints other than the spoken language contributed to the evolution of writing, such as the medium: clay and cuneiform in Mesopotamia, paper for Chinese ideograms. If the power of scribes and priests hampered the evolution of writing in many civilizations, we still owe the invention and use of the alphabet to the democratization of writing, mainly for commercial ends.

The Circuits of Writing

NEUROPSYCHOLOGY OF WRITING

The extreme concept of total genetic determinism in cerebral organization, which is still sometimes claimed, should in principle result in perfect reproducibility of brain anatomy in genetically identical individuals. But is it true? First, we must remember that in a genetically heterogeneous population there is remarkable individual variation in the topology of cortical areas in terms of cytoarchitectonics, as described by Korbinian Brodmann in 1909. This variation has been confirmed by fMRI in, for example, the topology of the visual areas excited by suitable visual stimuli. What is surprising is that there is even variability between monozygotic twins, notably in the speech areas. Helmuth Steinmetz and his colleagues selected ten pairs of monozygotic twins, both of whom in each case had the same (concordant) hand preference, and ten discordant pairs, consisting of one right-hander and one left. They observed that all the right-handers, whether from a concordant or discordant pair, showed asymmetry of the left *planum temporale* (part of the temporal cortex in which the human auditory cortex is located), whereas all the left-handers lacked this asymmetry. There thus seems to exist an epigenetic variability in cortical topology even in monozygotic twins, possibly due to an early difference in cleavage of the blastocyst during the first stages of embryonic development.

Since Broca's time we have known that a lesion of part of the left frontal lobe can result in speech deficits, or aphasia. In 1892 Joseph Dejerine described alexia, or pure word blindness (Figures 25, 26). Patients had severe difficulties with reading but not writing. They could write but could not read what they had written. In such cases Dejerine found a lesion, typically in

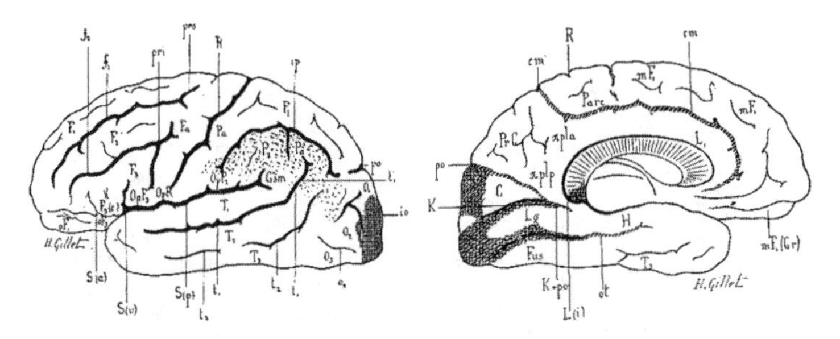

Figure 25: *Writing circuits according to Dejerine.* The area of the cortex marked with dots indicates the extent of a lesion caused by a cerebral vascular accident in a patient suffering from pure word blindness. From J. Dejerine, *Anatomie des centres nerveux.*

the white matter between the left angular gyrus and visual cortex. He concluded that there exists "a center for visual images of words." As alexia can coexist with agraphia, he postulated that this center is common to writing and reading, but that access to it from the visual centers can be altered selectively to result in pure alexia. Recent research has demonstrated that the reading pathways are more complex, and this overly simplistic model has been abandoned. We now know that peripheral alexia results from changes in the visual processing of words. For example, spelling alexia, which does not involve agraphia, results from left parietotemporal lesions and involves serious deficits in reading. The patient can read only letter by letter while spelling aloud, so the time needed to read a word depends on the number of letters in it. This is a selective change in input to the reading system before the separation of phonology and semantics. The lesion seems to be in a "visual word form system," according to Tim Shallice, and points to a neural mechanism common to alphabetic and ideographic writing that involves global processing of word

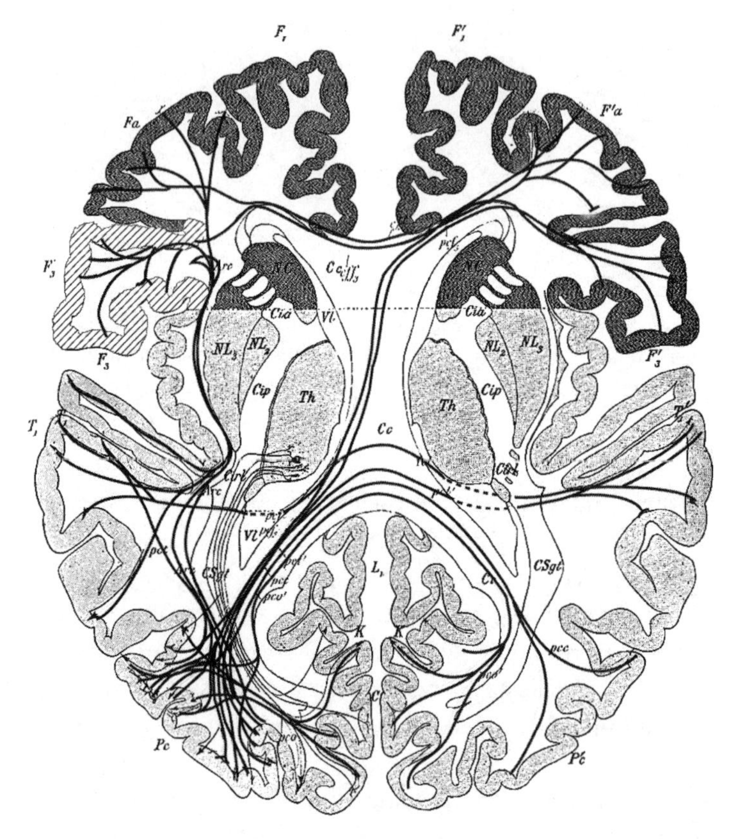

Figure 26: *Connections involved in written and spoken language.* There are strong connections between Broca's and Wernicke's areas and the frontal, motor, and visual cortex. There are also important commissural connections between hemispheres. From J. Dejerine, *Anatomie des centres nerveux.*

form. So reading would utilize a lexicon of visual signs learned in the course of learning to read. Other forms of peripheral alexia involve, for example, systematic errors, either at the beginning or the end of words (neglect alexia), or difficulty in reading lines of words in spite of being able to read individual words (attentional alexia).

On the other hand, central alexia involves processing after the visual analysis of written words. John Marshall and Freda Newcombe in 1973 distinguished two types of central alexia or dyslexia. Surface dyslexia involves selective deficits of reading of irregular words, such as *island* or *phase,* whereas regular words are read correctly. Parietotemporal lesions cause a deficit in relating a grapheme to a phoneme. An extreme case of phonological alexia was observed by Marie-France Beauvois and Jacqueline Dérouesné involving the selective reading of non-words (those not belonging to any visual lexicon) and grammatical words such as *and, if, for.* Deep dyslexia implies selective semantic deficits in reading words in relation to their meaning. For example, parietotemporal or frontal lesions lead to a deficit in reading either abstract or concrete words.

A model incorporating all these observations involves two parallel paths:

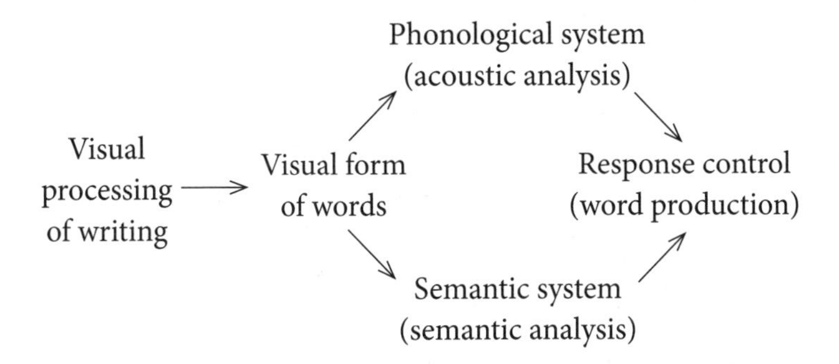

Just as alexia can exist without agraphia, so can agraphia without alexia. Certain discrete lesions interrupt the referencing of our memorized mental lexicon, which enables us to write unusual words correctly, as we already saw. Others disturb transcription of non-words, and still others the assembly of letters into words. There is apraxic agraphia, which prevents the voluntary production of signs, and other forms that disturb selection of upper- or lower-case letters. And there are others that interfere with orientation or spacing. Here again, two parallel paths come to mind:

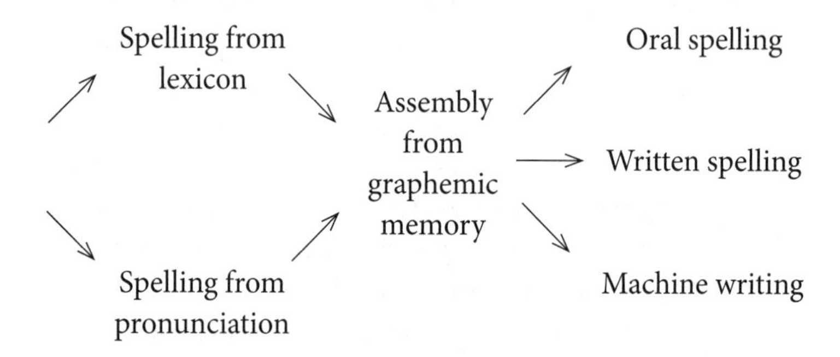

Japanese is a special case, in that there are two principal systems of writing: kanji, based on Chinese ideograms, and two forms of *kana,* comprising 71 syllabograms. Children learn kana at school before the age of six. Between six and fifteen they memorize about 1,000 kanji characters of the 3,000 in common use. In 1901 the neurologist Kinnosuke Miura described a patient with alexia and agraphia for kana, but with kanji intact. In 1984 Makoto Iwata distinguished three pathological conditions. Lesions of the left occipital lobe were associated with pure alexia for kanji and kana; the left angular gyrus with alexia for kana but not kanji, but with agraphia for both; and the left temporal lobe with alexia and agraphia for kanji, but the use of kana was unaffected. To explain these results, he proposed two distinct pathways for reading Japanese, a dorsal for kana and a ventral for kanji.

The connections of the brains of illiterate people differ considerably from those of adults who learned to read and write in childhood. Alexandre Castro-Caldas from Portugal and Martin Ingvar from Sweden used PET scans to compare regional blood flow in twelve women aged sixty-three to sixty-five who were from similar sociocultural backgrounds and of similar cognitive ability; some were illiterate and others were able to read and write. It was already known that illiterate subjects present selective difficulties in repeating from memory any non-words in a list of normal words. In experimental conditions in which subjects have to repeat words and non-words, there is left inferoparietal-temporal dominance. The difference between words and non-words reveals activation of the fronto-opercular cortex, right anterior insular cortex, and left anterior cingulate cortex. Learning to read is accompanied by stabilization of cerebral circuits that are involved in the phonological processing of new words. There is therefore considerable interaction between the circuits for written and spoken language; the acquisition of written language significantly modifies this interaction.

Functional imagery of reading has revealed preliminary results suggesting a differential contribution of cortical areas: primary and secondary visual areas for analysis of the form of words, Broca's area for phonological coding of non-words, the left superior temporal area for rhymes and phonological articulation, and the left prefrontal and anterior cingulate for semantic tasks. There exist therefore several forms of plasticity in the cerebral circuits for reading, with very different temporal relations: attention is measured in milliseconds, practice in seconds or minutes, learning in hours or days, deduction of rules in weeks or months, and development in months or years. Writing circuits are established by selective stabilization of connections provided by the genetic envelope through prolonged interactions with a child's cultural environment. Learning to write corresponds to recruitment of preexisting immature neuronal circuits. Cultural circuits are imprinted in the child's brain for life.

IV
The Molecular Biology of the Brain

Genes and Phylogenesis

THE QUESTION OF ORIGINS: CREATIONISM OR EVOLUTION

Before science developed, humans elaborated mythical reflections to give some meaning to events and experiences they encountered and to establish classifications that were "superior to chaos," as Claude Lévi-Strauss said. Theories included creation by gods and spontaneous generation. Common concepts included great floods to punish sins and re-create the world, and creation of life in a primeval ocean by successive steps. In the West the myth of dualism of body and soul emerged. The dualist, creationist view was opposed from ancient Greece onward by more materialist concepts, some of which emphasized that diverse elements could be combined randomly to form living creatures. Democritus saw the world as composed of atoms in constant transformation and combining randomly to form organisms. These concepts are not far removed from the Darwinian and Lamarckian approaches, which we have already discussed.

The founding fathers of modern evolutionary theory can be found among the philosophers of the European Enlightenment. Maupertuis adopted an Epicurean notion of mutation and elimination of defective variants. Buffon envisaged the possibility of modern species descending from common ancestors and even suggested that humans and monkeys had a common origin, an idea for which Lucilio Vanini was burned at the stake by the Inquisition a century earlier. Buffon classified species in terms of their region of origin, their behavior, and their internal anatomy and so founded biogeography, ethology, and comparative anatomy. He insisted, however, on the immutability of species. On the other hand, for Diderot the living world was "a machine that advanced toward perfection by an infinity of successive developments." He had an authentic evolutionist view of a world without a plan, made of continually modifying molecules with randomly appearing forms, of which the most resistant survived. He extended his concepts to the development of ideas: "the advancement of the spirit is merely a series of experiments" by "imaginary beings," which the brain "combines"; he contended that reasoning occurs "through successive identifications."

COMPARATIVE ANATOMY AND PALEONTOLOGY OF THE BRAIN

The first forms of life, probably prokaryotic photosynthetic bacteria, appeared on the earth more than three billion years ago. The first eukaryotic organisms, which contained a true nucleus, were the green algae, about one billion years ago. The first multicellular organisms with a nervous system (coelenterates, polychaete worms, soft arthropods, and specimens of extinct phyla) left evidence in the form of burrows, tunnels, or tracks from about 640 million years ago, the Ediacaran fauna. In Cambrian sediments from 590 million years ago we already find fossils of all phyla that had fossilizable skeletons, including the first chordates, such as *Pikaia.* Vertebrates appeared with the armored fish or ostracoderms in sediments aged 500 million years. The most primitive unicellular eukaryotes, or protists, already used chemical signals for intercellular communication, such as α sex factor in yeast, β-endorphins in *Tetrahymena,* or cyclic adenosine monophosphate (AMP) in *Dictyostelium,* as well as their receptors. They had ion channels for calcium and potassium in their membranes, and they produced calcium action potentials. Voltage-gated sodium channels were not yet present; they appeared only with multicellular organisms.

A sponge has a body consisting of a jellylike mass between two epithelial layers. Sponges exhibit multicellular coordination—for example, rhythmic contractions every thirty seconds at most, without the intervention of real nerve or muscle cells. Max Pavans de Ceccatty noted the presence of primitive junctions or synapses between their cells, though these did not attain the level of differentiation found in higher animals. In cnidarians, *Polypodium,* and jellyfish, there exist well-structured neurons and synapses. True action potentials with voltage-gated sodium channels have been recorded in the jellyfish *Aglantha,* as well as a curious phenomenon of bidirectional chemical synaptic transmission. All the biogenic amines and neuropeptides of the *FMRFamide* type are present, but not acetylcholine. In *Aglantha* primitive forms of long-range nonsynaptic chemical transmission, called volume transmission by Kjell Fuxe, are found, accompanied by non-quantum release of neurotransmitter. They also have giant axons propagating fast sodium action potentials for escape, whereas slow, rhythmic swimming behavior utilizes much slower calcium potentials in the same axons. It is remarkable that the very primitive nervous system of these coelenterates is already organized around endogenous spontaneous activity modulated by interaction with the outside world. This very general functional principle is quite distinct from that of the instructional cybernetic concept, which modulates behavior in terms of input-output relations.

In higher animals, the triploblasts (which have three embryonic layers and a coelom, or body cavity), the nervous system follows a common organizational plan in spite of large anatomical variation. Etienne Serres noted this unity as early as 1826 in invertebrates and extended it to vertebrates. The principal homology resides in the segmental organization of the body, which features repeated segments, or metameres, and the fusion of certain of them to form the head and brain. The invertebrate nervous system is composed of compact ganglia, whereas the vertebrate nervous system develops from a hollow neural tube and has the considerable geometric advantage of being able to grow easily by surface proliferation. Furthermore, in invertebrates the cerebral ganglia are dorsal (along the "back" of the body) and the main nerve chain is ventral; in vertebrates the brain and spinal cord are both dorsal. Alfred Romer proposed several mechanisms to account for the change from the invertebrate type of nervous system to the vertebrate type: a dorsal growth of the spinal cord from the cerebral ganglia, a dorsoventral inversion (vertebrates are said to be annelid worms walking on their backs), or a partial change of segmental dorsoventral polarity. In 1828 von Baer described the segmentation of the vertebrate central nervous system into neuromeres. In most vertebrates the embryonic neural tube subdivides into three vesicles—anterior, middle, and posterior—from which derive the forebrain, midbrain, and hindbrain; the forebrain is further divided into telencephalon and diencephalon. In 1969 Vaage distinguished eight neuromeres in the forebrain of the chicken, two in the midbrain, and seven or eight in the hindbrain.

The evolution of the vertebrate brain takes place by differential growth of size and complexity of each segment. Overall the weight of the brain increases over that of the body by a power of 2/3. In fact, it is more proportional to the body surface area than its weight. The differences in brain weight for species of the same biological group but of different body size can be plotted on a straight line. We can define an encephalization index equal to brain weight/body weight as 2/3. So we can define a progression index as the ratio of the encephalization index of a given species compared with that of a basic insectivore, considered as a unit. The progression indexes of prosimians, orangutans, chimpanzees, and humans are about 4, 9, 11, and 29. Progressive encephalization marks the course of vertebrate evolution, but in parallel we find corticalization. The oldest paleocortex is found in cyclostomes up to mammals, in which it becomes the pyriform or olfactory lobe. Next comes the archicortex, originally dorsal but now internalized to form the hippocampus. The more recent neocortex developed explosively from reptiles to higher mammals to produce the typical six layers of superimposed neurons. Two rules appear to control the evolution of the neocortex: a change from a radial, three-layered to a six-layered organization, and tangential growth in area accompanied by an increase in vertical columns. Data from human pathology, especially that related to congenital deficits, supports these concepts. Microcephaly is associated with a reduction in the number of columns: microcephalia vera is a deficit in transverse cell proliferation, and Zellweger syndrome involves disturbed migration of neurons. In primates the visual system is highly developed and ensures good binocular vision, but the olfactory system is reduced. The brain increases in weight with growth of the surface area of the neocortex, which takes on a complex form of depressed furrows called sulci between raised convolutions called gyri. Certain cortical areas specialize and increase in relative size, such as visual areas as well as the temporal and frontal cortex. The cerebellum, involved in visuomotor control, also grows in complexity.

Hominization first appeared as bipedalism. Humans are the only living primates to be truly bipedal. The human pelvis differs from that of the chimpanzee by the angle between the ischium and the ilium, which enables development of the gluteal muscles necessary to maintain upright posture. There is also an increased volume of the pelvic cavity to allow the passage of the large head of the human fetus. The pelvis of *Australopithecus,* who was already bipedal, was similar to that of modern man. So bipedalism preceded the expansion of the brain of humans' direct ancestors. The foot decreased in relative length, and the tarsus lengthened with respect to the toes, thus offering an effective lever for walking. The big toe was no longer opposable to the other toes: it lengthened and was able to support half the body weight. The hand remained primitive, but the fingers, as well as the joints of the arm, became more flexible. The fingers became flatter and offered better touch sensation. The hand replaced the jaws for prehension, so the jaws and the face became smaller, allowing the larynx and pharynx to descend and the tongue to be freed, so preparing for articulated language. The bony orbit became finer and the eye movements more independent. The foramen magnum, the hole in the base of the skull through which the brainstem becomes continuous with the spinal cord, was lowered, and the occipital condyles, the bony prominences that articulate the skull on the first neck vertebra, advanced, favoring expansion of the cranium. During the last five million years the size of the brain has at least tripled. According to Harry Jerison, this increase was strictly allometric, that is to say, in constant relationship to body weight. But for Ralph Holloway a heterogeneous selective reorganization of the brain accompanied its relative growth: there was a "mosaic" evolution. The progression index of the neocortex increased from 10 to 60 from the lemur to the chimpanzee and jumped to 156 in modern humans. From chimpanzees to humans the primary sensory cortex decreased in relative area, whereas that of the association cortex (which does not receive direct afferent inputs from sensory systems or control motor activities, at least directly) increased. The prefrontal association cortex occupies 11 percent of the neocortex in macaques , 17 percent in chimpanzees, and 29 percent in humans.

Terrence Deacon (1988) attempted to find language circuits in the monkey. He found circuits in the diencephalic and midbrain limbic systems involved in vocalization, but, being subcortical, they presented no obvious homology with human cortical language areas. On the other hand, stimulation of the inferior limb of the arcuate sulcus, the homologue of Broca's area, caused contraction of facial, oral, and laryngeal muscles. Similarly, area Tpt, a homologue of Wernicke's area (in the temporal cortex and, like Broca's area, related to speech), intervened in discrimination between vocalizations produced by the individual and other members of the same species (Figure 27). Deacon established a map of connections between these areas in monkeys and noted their conservation from monkeys to humans. In his opinion, language areas appeared not de novo but by quantitative remodeling and functional reorientation of preexisting circuits. Further, Deacon compared the progression index of laminated structures, such as the cerebral cortex and cerebellum, of nonhuman primates with that of nuclear structures in the striatum, diencephalon, and medulla and demonstrated that the former grow much more rapidly than the latter.

By 1836 Etienne Geoffroy Saint-Hilaire had already noted that the skull of the young orangutan strikingly resembled that of a human child, whereas adult skulls of the two species were very different. In 1926 Louis Bolk considered that numerous characteristics of the human adult were fetal characteristics that had become permanent: what was a transitional stage in ontogenesis in other primates had become a terminal stage in humans. Bolk cited, for example, orthognathy (that is, the verticalization of the jaws) to give a vertical aspect to the face, reduced body hair and pigmentation, the shape of the ear, the mongoloid eye, the position of the foramen magnum, the high relative weight of the brain, the structure of the hand and foot, the shape of the pelvis, and the orientation of the vagina. For Stephen Jay Gould "a general delay of development characterizes human evolution. . . . This retardation established a matrix within which all trends in the evolution of human morphology must be assessed." Although human gestation time is close to that of anthropoid apes, the human brain, on the contrary, continues to grow in size and complexity long after birth. Human cranial capacity at birth is 23 percent of the adult capacity; compare this to 40 percent in the chimpanzee and 65

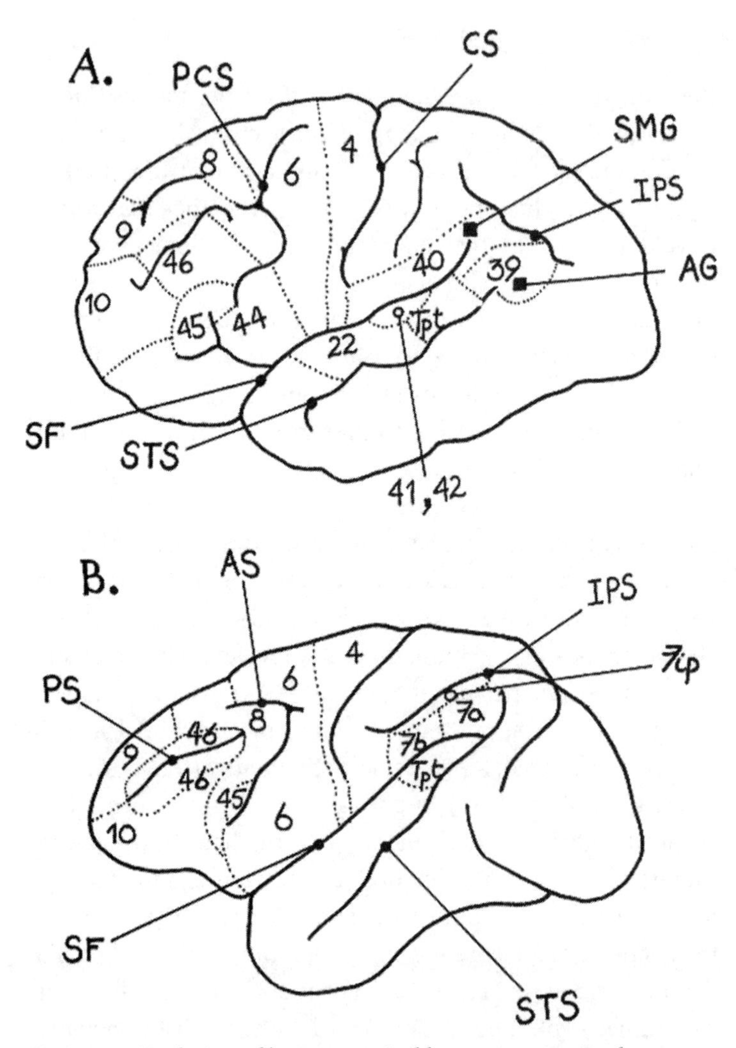

Figure 27: *Evolution of human cortical language areas.* In the cortex of the macaque (lower drawing) we find homologues of human (upper drawing) language areas, such as area 45, perhaps homologous to the human Broca's area (44, 45), and Tpt, perhaps homologous to Wernicke's area. From F. Aboitiz and R. García, "The Evolutionary Origin of the Language Areas in the Human Brain: A Neuroanatomical Perspective," *Brain Research Reviews* 25 (1997): 381–396; reprinted with permission from Elsevier.

percent in the macaque. In other words, prolonged brain development together with precocious birth allows lengthy interaction with the outside world and thus more efficient neurocultural mediation.

Paleontological records of the brains of our ancestors are very incomplete. *Dryopithecus,* who lived between 19 and 10 million years ago, and *Ramapithecus,* between 14 and 10 million years ago, are believed to be our most distant hominid ancestors, having diverged from modern anthropoids 8 million years ago in Africa. Chimpanzees are said to have survived in the western tropical forests. According to Yves Coppens, hominids appeared in the east in the drier savanna of the high plateaus. *Australopithecus* was present between 5 and 3 million years ago. Their cranial capacity was 400 milliliters. Between 2.5 and 1 million years ago they coexisted with *Homo habilis* (700 milliliters). *Homo erectus* (800 to 1,000 milliliters) developed 1.5 to 0.5 million years ago and gradually gave way to *Homo sapiens neanderthalensis* (1,600 milliliters), then to *Homo sapiens sapiens* (1,000 to 2,000 milliliters) 100,000 years ago. According to Holloway, the brain of *Australopithecus afarensis* had decreased primary visual cortex and increased visual association areas, notably in the inferior parietal cortex, which was specialized for multimodal visual processing. In *Homo habilis* the frontal lobe, notably Broca's area, was reorganized, and there was an increase in cerebral asymmetry. We do not know if *Homo habilis* could speak. *Homo erectus* underwent an allometric increase in brain size, which continued with *Homo sapiens,* in whom we see additional complexity and non-allometric asymmetry.

Tools—lithic cores, flakes, and broken flints—have been found with remains of *Australopithecus gracilis,* 2 or 3 million years old. *Homo habilis* used choppers. He constructed huts and undertook cooperative activities, such as hunting, gathering, and transport; he shared food and educated his children. He also fabricated secondary tools using primary tools. *Homo erectus* used bifaces (two-sides stone tools) and lived in mobile camps and shared the workload. He domesticated fire 1.4 million years ago in Kenya and 750,000 years ago at Escale in France: he diversified his cultures. *Homo sapiens* is associated with sophisticated Mousterian tools, such as flint, bone, or ivory bifaces, arrowheads, and scrapers. The first burials and funeral rites emerged. Religious cults developed, some perhaps associated with cannibalism. In certain sites *Homo sapiens neanderthalensis* and *Homo sapiens* shared the same culture at the same time merely a few kilometers apart. Art emerged in Europe and Australia some 35,000 years ago. The Neolithic revolution took place 10,000 years ago, and man became sedentary. In the early stages of hominization cultural evolution developed in relation to biological evolution; in later stages that was no longer the case.

GENETIC EVOLUTION AND HOMINIZATION

The human genome, carried in the DNA (deoxyribonucleic acid) of our forty-six chromosomes, comprises 3×10^9 base pairs and several tens of thousands of structural genes, a large number of which are expressed in the brain. Only some 3 to 7 percent of the human genome codes for structural genes; a small percentage is concerned with regulation, and the rest have no known function. There are numerous homologies among the structural genes, probably due to multiple genetic duplications. Among the resultant families of genes we find those for neurotransmitter receptors, but also transcription factors, which are expressed during embryonic development and contribute to the gene regulatory scaffold that precedes the expression of the phenotype.

From chimpanzees to humans the mean differences in sequence between proteins is no more than 0.8 percent. Between genomes this is around 1.1 percent, according to Mary-Claire King and Allan Wilson (1975). Human chromosomes differ from those of the chimpanzee notably by nine pericentric inversions, the fusion of chromosomes 2p and 2q, and a deletion in chromosome 13. The chimpanzee is most closely related to humans, but the gorilla is also very close, the orangutan remoter, and the gibbon even remoter. The genetic distance between different living human populations represents between 1/25 and 1/60 of the distance separating chimpanzees and humans. Mitochondrial DNA is transmitted from mother to daughter; it enabled Allan Wilson to construct a phylogenetic tree of which all the branches led to a common African ancestor. This "Black Eve" would have lived between 280,000 and 140,000 years ago. Would she have been *Homo sapiens* or *Homo erectus*?

Darwin insisted on the continuity of evolution (*natura non facit saltum*), as opposed to the discontinuity of the creationists, for whom man appeared suddenly by divine intervention. William Bateson (1861–1926) and Hugo de Vries (1848–1935) soon introduced the concept of discrete mutations of the genome, however, to explain the origin of the diversity of species. Indeed, the very definition of species includes the notion of discontinuity. Ernst Mayr (1904–2005) defined a species as a reproductive community, a mutually fertile natural population, reproductively isolated from other similar groups, an ecological unit that interacts with other species sharing the same environment, and a genetic unit formed from an intercommunicating gene pool. According to him, segregation into species prevented a too great genetic variability; species were the real units of evolution, each realizing an evolutionary experiment that possessed limited ecological significance but was able to compete with others for a given environment. Segregation of new species was produced by geographical isolation. The process of random sampling of small or very small populations within a larger heterogeneous population was accompanied by a genetic drift with a founding effect able ultimately to cause massive genetic changes.

In 1944 George Simpson noted considerable differences in the speed of evolution in different groups and species. A living species of crustacean, *Triops cancriformis*, has remained the same for 180 million years, whereas it had taken only three million years for *Australopithecus* to become human. Gould took up Simpson's point and showed that in the same lineage, after often very long periods of stasis, rapid changes followed each other (punctuated equilibrium) that could be explained by Mayr's "founding" populations. Small populations were subject to the fastest genetic changes, and it seemed plausible that the transition from *Australopithecus* to *Homo habilis* corresponded to a "punctuation," but the change from *Homo habilis* to *Homo sapiens* seems to have been more gradual. In 1920 Richard Goldschmidt proposed the idea that gene mutations act on early embryonic development, accomplishing in a single stage a considerable evolutionary point of departure by the production of what he called "hopeful monsters." In 1929 two Soviet scientists isolated mutant *Drosophila* in which homologous appendages grew on segments to which they did not belong (for example, a leg instead of an antenna). The genes involved were called homeotic, and their potential importance in macroevolution was underlined by Goldschmidt in 1940 and by Simpson in 1944. Recent analysis of these genes by Edward Lewis and Walter Gehring, along with that of several families of development genes, has led to the formulation of original hypotheses about their contribution to the evolution and development of the nervous system in vertebrates.

HUMAN NATURE *IN SILICO*

In 2001 the journals *Nature* and *Science* announced the initial, although incomplete, sequencing of the human genome. Human nature was finally deposited *in silico* in the form of nucleic sequences. The two rival teams, each of around 300 persons, reported similar results, but there nevertheless were substantial differences. These results followed earlier work to completely sequence the genomes of unicellular organisms such as *Escherichia coli* and the yeast *Saccaromyces cerevisiae* (1996), the multicellular nematode worm *Caenorhabditis elegans* (1998), the fruit fly *Drosophila melanogaster* (2000), and the flowering crucifer *Arabidopsis thaliana* (2000). The comparative analysis of these genomes was then possible. As we might expect, the size of the genome grew with evolution: from 4.7 million bases (Mb) for *E. coli* to 13.5 for yeast, 100 for *C. elegans,* 165 for *Drosophila,* and 2,910 for humans. In plants the genome is generally large, from 120 Mb for *Arabidopsis* to 2,500 for maize and even 16,000 for wheat. On the other hand, the number of genes evolves less expansively than the number of bases, from around 400 for *E. coli,* to 6,000 for yeast, 13,000 for *Drosophila,* 18,000 for *C. elegans,* and 25,000 for *Arabidopsis.* The number of human genes coding for protein sequences is still uncertain: more than 39,000 for Craig Venter's Celera Corporation; 32,000, according to the International Sequencing Consortium; and, more recently, suggestions are that it might be still fewer, around 20,000. These values are much less than the 100,000 initially suggested by Walter Gilbert and others.

The percentage of bases occupied by genes varies around

1 percent for coding sequences (*exons*), about 30 percent for *introns* (noncoding sequences intercalated in a structural gene), and about 70 percent for noncoding sequences that separate the genes. A very small fraction, less than 1.5 percent, of the total DNA sequences of the human genome effectively codes for proteins. The noncoding sequences consist essentially of repeats (about 50 percent of the genome). For the International Sequencing Consortium these repeats are not "junk" but an extraordinary source of information: passive markers of paleontological history, as well as active agents of change in the form of the genome. For example, in the human genome about forty-seven genes are derived from retrotransplantation, that is, insertions resulting from reverse transcription of RNA (ribonucleic acid) to DNA. We should recall that duplications among and within chromosomes can cause human genetic disease such as *adrenoleucodystrophy* and Charcot-Marie-Tooth disease. Transposons are DNA sequences that can move to different positions in the genome by transposition. Their activity was very effective in the human lineage 150 to 80 million years ago, but stopped 50 million years ago. A single transposon has intervened in the human genome since our divergence from the chimpanzee, and it is not even shared by the whole human race. In the mouse 10 percent of mutations result from transpositions; the proportion is less than 0.2 percent in humans. The reason for this lower frequency in man is not known. It may be due to the small size of the ancestral population at the origin of human speciation. Introns form another important category of noncoding sequences in the human genome. Their size increased during evolution from invertebrates to humans, whereas that of exons did not change: there is perhaps a link with the increase in the number of genes presenting alternative splicing, 22 percent in *C. elegans,*

35 percent in humans. The size of introns increases in regions of the genome poor in guanine-cytosine, where the distance between genes also increases.

Only about 60 percent of the coding sequences that translate into proteins, the human proteome, had been identified in 2001. The functions of the approximately 30,000 known genes include nucleic acid binding (13.5 percent), signal transduction (12 percent, of which 5 percent are for receptors), and those related to enzymes (10 percent). We must distinguish some of these proteins. Housekeeping proteins are homologues (*orthologues*) of proteins present in the genome of a fly (61 percent), yeast (46 percent), and a worm (43 percent) and form the heart of the human proteome, of which over 1,500 proteins are common to man and invertebrates. As their name indicates, these proteins ensure basic functions of cell life such as division, protein synthesis, and glycolysis. Multicellularity proteins distinguish multicellular organisms, such as *C. elegans,* from unicellular ones, such as yeast. They contain novel protein domains and considerably expanded existing domains, which serve to form cell assemblies. Among them we find proteins involved in intercellular and intracellular signal transduction (such as epidermal growth factor—EGF—catenin, and phosphotyrosine), intercellular adhesion (fibronectin), apoptosis (cell death: caspases), transcription (hormone receptors, homeotic proteins), and protein-protein interactions. The distribution of multicellularity genes on the chromosomes in *C. elegans* shows a privileged distribution of the most recent genes on their arms rather than at their center. Could this be an argument in favor of Ernst Haeckel's (1834–1919) terminal addition theory? A remarkable phenomenon is that many genes responsible for human neurological disease, such as amyotrophic lateral sclerosis, hereditary deafness, anencephaly, and fragile X syndrome, are orthologues of genes that exist in flies, worms, and even yeast.

From invertebrates to humans, proteins develop that are implicated in defense and immunity, transcription and the cytoskeleton. Only 7 percent of gene families are specific to vertebrates and humans; that is 94 families, of which 23 are involved with defense and immunity and 17 are specific to the nervous system. New protein structures are in fact derived by reassortment, addition, or deletion of old protein domains. There has been evolution by domain accretion with fusion at the extremities of existing domains. Invertebrate genes have multiplied with changes of function: for example, immunoglobulin domains present as surface proteins in invertebrates have been reoriented for the production of antibodies in vertebrates. Certain gene-rich families, such as olfactory receptors (906 receptors and pseudogenes), largely lost their function in the human lineage in the last 10 million years. On the other hand, certain transcription factors multiplied from flies to humans (90 to 220 homeogenes); new domains appeared, such as KRAB/SCAN sequences, which may intervene in the assembly of transcription factors into oligomers that form transcriptional networks.

The increase in complexity of the nervous system during evolution was accompanied by significant changes in the genome: the "proteome of neuronal man." The number of genes coding for the propagation of neural activity has increased. Eleven human genes code for the sodium channel, versus four in the fly and none in the worm. The number of genes for transmitter receptors with seven transmembrane domains increased from 146 in the fly and 284 in the worm to 616 in the human. Some genes are found exclusively in vertebrates, including humans: myelin (10), connexins (14), opiates (3), and calcitonin gene-related peptide (CGRP) (3). Cytoskeletal genes developed massively from worm to fly to human (actin 12 to 15 to 61, annexin 4 to 4 to 16, spectrin 10 to 13 to 31). Genes involved in the development of synaptic connections also proliferated: genes for nerve growth factor (NGF) and neuroregulins exist only in man. Genes for fibroblast growth factor (FGF) increased from 1 to 1 to 33 from worm to fly to man, for ephrins from 4 to 2 to 7, for their receptors from 1 to 2 to 12, for cadherins from 16 to 14 to 100, and for semaphorins from 2 to 6 to 22. Finally, transcription factors involved in morphogenesis increased spectacularly: for example, the number of so-called CZH2 zinc finger proteins grew from 234 in the fly to 4,500 in man. The number of cell death proteins, such as caspases, is 0 in yeast and only 2 in the worm, but 16 in man.

The development of the proteome associated with the development of the nervous system corresponds mainly to increased complexity of the network of connections. The unraveling of the genome, however, revealed the parsimony of human structural genes and also the nonlinear small number of genetic changes responsible for the enormous increase in complexity of the nervous system. We may understand this apparent paradox if we define, as Jean-Michel Claverie did, the complexity of an organism as the theoretical number of possible states of its transcriptome. If there exist only two possible ON or OFF states for each gene, such that N genes code 2^N states, human genetic complexity would be some $2^{20,000}$, a very large number, far larger than the total number of particles in the universe. Thus, there is no serious theoretical limit to the number of possible combinations. This is even truer if we suppose that states of genetic expression are graduated rather than all or nothing. The problem is rather the inverse: how to master such a massive combinatory explosion and ensure a sufficiently robust organism to resist natural selection.

GENETICS OF BODY SHAPE: THE EXAMPLE OF *DROSOPHILA*

It is the biologist's job to identify molecular mechanisms for determination of the organization of the body. In *Drosophila* several gene families specify the body plan, including the nervous system. Many are what Goldschmidt termed homeotic and which were labeled *homeobox* by Gehring. As Christiane Nüsslein-Volhard demonstrated, the anteroposterior axis of the body is controlled by maternal effect genes, bicoid for the anterior and oskar for the posterior part of the body, which code for homeobox proteins. The primary subdivision of the embryo into head, thorax, and abdomen is determined by gap zygotic genes (hunchback, Krüppel, knirps), which code for zinc finger proteins and in which mutations cause elimination of certain body parts. The segmental divisions of the embryo, the metameres, are controlled by pair-rule genes (such as runt, hairy, fushi-tarazu) and their polarity (for example, engrailed and wingless). The identity of each type of head, thoracic, or abdominal segment is specified by true homeotic genes (antennapedia, bithorax). Finally, the dorsoventral axis is determined by maternal effect genes (snake, dorsal, Toll, among many others). The central nervous system is derived from cells of the ventral ectoderm under the control of dorsoventral polarity genes (single-minded) or neurogenic genes (notch), which limit the entry of ectodermal cells into the neuronal line. The genes that are expressed first are in general for transcription factors, and the last ones code for receptor proteins, G proteins, and membrane proteins involved in intercellular signaling. Early genes are expressed again later in the nervous system, which is the case, for example, for fushi-tarazu.

The idea of a conserved plan of organization of the body and nervous system from invertebrates to vertebrates was reinforced by the identification in vertebrates of developmental genes homologous to those of *Drosophila.* A homeodomain sequence helped identify in mice, and then in humans, a family of four groups of Hox genes, homologues of antennapedia in *Drosophila.* Other homologous families of engrailed, even-skipped, caudal, and paired genes of *Drosophila* have been isolated in the mouse. Many of these genes are expressed segmentally in the central nervous system, such as Hox 2.1, 2.6, 2.7, 2.8, 2.9, Int2, and Krox 20. Inactivation by homologous recombination of Hox 1.5 produces a segmental deficit restricted to the region of the thyroid; that of Int1, homologue of winglessness in *Drosophila,* causes a deficit in the midbrain and cerebellum. We have thus every reason to believe that genes of this sort determine the development of the nervous system in both invertebrates and vertebrates (Figure 28). These genes can undergo heterochronic mutations, which lead to differences in their temporal expression or in the expression of genes that they control. Several have been identified in *C. elegans:* they cause either acceleration or delay in development as evaluated by the number of cell divisions. The difference between the neotenic *Amblystoma mexicanum* in its aquatic larval form and *A. tigrinum,* which produces terrestrial salamanders, is, for example, due to the mutation of a control gene probably involved in hypothalamic or hypophyseal stimulation.

A plausible genetic model of vertebrate evolution can therefore be proposed, controlled by several gene types. Differential action on body segments of dorsoventral polarity genes of the Toll type produces a dorsal organization of the nervous

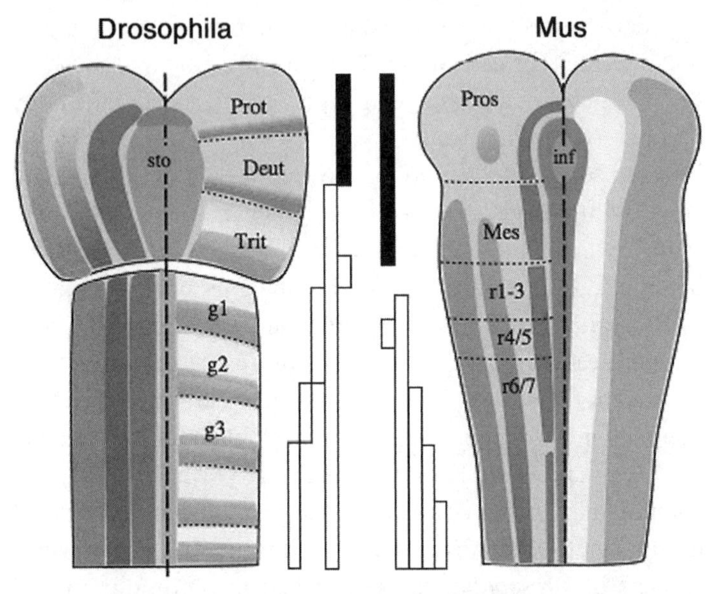

Figure 28: *Analogous development of the nervous system of* Drosophila *and mouse (Mus).* Comparison of various developmental genes at similar stages of the neural tube, here seen open as the neural plate. Note the overall similarity between the two species. From D. Arendt and K. Nübler-Jung, "Comparison of Early Nerve Cord Development in Insects and Vertebrates," *Development* 126 (1999): 2309–2325; reproduced with permission of *Development.*

system in vertebrates or a ventral organization in invertebrates. Other genes determine the formation of a hollow neural tube in vertebrates or a solid chain of ganglia in invertebrates. Activity of segmentation or homeotic genes causes parcellation and differential expansion of cortical areas by radial and tangential cellular proliferation. The telencephalon comprises neuromeres pr1 and pr2 in the chicken but could contain at least five neuromeres in mammals that coincide with the rudimentary cortex of the paleo-, archi-, and neopallium. Genes determining the organization of these last three neuromeres could have acted in a privileged fashion in the evolution of the brain from *Australopithecus* to modern humans.

CONSEQUENCES OF RESULTS FROM *DROSOPHILA*

The fact that important genetic homologies exist between *Drosophila* and vertebrates does not signify that homologous genetic elements are expressed in exactly the same ways during development in different animal groups. In *Drosophila* the spatial coordinates of the embryo are determined before fertilization by the expression of dicephalic or Toll genes during oogenesis, the process of formation and maturation of the egg. The two relevant symmetry breaks are found at the level of the topological relationships of the oocyte ("egg cell") with its supporting cells. The egg of *Drosophila* is predetermined and mosaic. The egg of the amphibian is different. Although it is laid with an anteroposterior polarity, indicated by a pigmented animal pole that corresponds to its attachment to the ovary, it does not have dorsoventral polarity. This is determined by the site of penetration of the sperm, which under normal gravitational conditions marks its ventral aspect and thus its bilateral symmetry. The first symmetry break results from the attachment of the oocyte to the ovary and the second from the entry of the sperm (Paul Ancel and P. Vitemberger, 1948, and, more recently, John Gerhart).

The mammalian egg represents a more extreme situation. It seems perfectly isotropic after fertilization, until stage 8. From stage 8 to 16 two cell types differentiate. Internal cells become the embryo, and external cells the trophectoderm. A first symmetry break appears at stage 32, when the inner cell mass detaches from part of the trophectoderm to form the *blastocele.* The attachment of the inner cell mass to the trophectoderm defines the dorsoventral polarity of the embryo. Anteroposterior polarity follows. L. J. Smith (1985) studied the orientation of the spatial coordinates of the mouse embryo four and a half days after implantation in relation to those of the uterus. This orientation seemed statistically fixed, leading to the proposal that the uterine environment of the embryo, in terms of tissues and hormones, for example, determines its spatial coordinates.

Whatever the case, the two critical fundamental symmetry breaks that determine the spatial coordinates of the embryo are different in different animal groups: they occur during oogenesis in *Drosophila,* at the end of oogenesis and at fertilization in amphibia, and after fertilization in the mouse. Nevertheless, the possibility that homologous genetic determinants are involved, and that the differences may be due to the process of distribution or redistribution of resulting cytoplasmic factors, cannot be excluded. A way of testing this hypothesis would be to seek homologues of dicephalic, Toll, and dorsal in vertebrates. The development of an egg seems to involve temporal and spatial deployment of a dynamic flux of genetic interactions, including convergence and divergence of regulating signals and reemployment of genetic determinants. There is both epigenesis by differential expression of genes and preformation by stable coding of genetic information at the DNA level and of certain gene products in the cytoplasm.

TURING AND BRAIN EVOLUTION

Following a long historical tradition going back at least to Jacques Loeb's *Mechanistic Conception of Life* (1912) and Joseph

Needham's *Chemical Embryology* (1929), our modern view of theoretical and experimental research on embryonic epigenesis is based mainly, explicitly or not, on Alan Turing's theory of 1952 in his "Chemical Basis of Morphogenesis": "It is suggested that a system of chemical substances, called *morphogens*, reacting together and diffusing through a tissue, is adequate to account for the main phenomena of morphogenesis." It is, however, necessary to have interaction between chemical reactions, with autocatalysis and feedback, giving rise to nonlinear processes with symmetry breaks. Embryonic development results from progressive and concerted differential activation or inactivation, in space and time, of the 20,000 or so human genes. A network of intergenic communication with local autocatalysis and long-distance inhibition, suggested by Hans Meinhardt and Alfred Gierer in 1974, would provide the conditions required to create symmetry breaks. The hypothesis seems plausible because forms of diffusion have been recognized in the first stages of development of *Drosophila* by Nüsslein-Volhard, and also in vertebrates, directly or indirectly and with signal relay at the membrane level, by Kerszberg and Changeux, 1998, as well as Alain Prochiantz. Such a network could be conceived at the level of transcription factors. These regulating proteins bind to sequences specialized for regulation, or promoters, first identified by François Jacob and Jacques Monod in 1961 and usually located upstream of the coding part of structural genes.

The hypothesis offered by Kerszberg and Changeux in 1994 was that the network is constructed by interaction between diffusible transcription factors assembled into several types of oligomer that are read differentially by promoter sequences. The simplest example is at the neuromuscular junction, where there is a virtual boundary of gene expression limited to the muscle fiber directly below the nerve ending. The postulated genetic communication involves, among others, a MyoD type transcription factor that regulates its own transcription at the level of the promoter of the gene that coded for it. A more complex example is the reading of a morphogenetic gradient that positions a band of gene expression during embryonic development. According to this model, a diffusible morphogen of maternal origin is assembled as a hetero-oligomer containing a transcription factor of embryonic or zygotic origin. There then forms a sort of "molecular synapse" between a gene product initially expressed in the egg and another expressed by the developing embryo. The activity of the hetero-oligomer on the transcription of the zygotic factor differs from that of the homo-oligomers resulting from the polymerization of one or another of these components. Computer simulation of the model predicted the positioning of a band along the anteroposterior axis of the embryo and its displacement as a function of minor changes (for example, of the gradient of diffusible morphogen) (Figure 29). Later, in 1998, Kerszberg and Changeux proposed a model for the formation of the neural plate by interaction among transcription factors and by genetic control of intercellular adhesion and cell movement leading either to invagination to form a neural tube or to detachment of neuroblasts and their compacting into a chain of ganglia.

These different models might explain two important aspects of the evolution of the central nervous system, the formation of a hollow neural tube in vertebrates or a chain of ganglia in invertebrates, but also the differential expansion and parcellation of the cerebral cortex in mammals. The data of Robert Barton and Paul Harvey in 2000 comparing the volume of the cortex in insectivores and primitive primates could be explained by a mosaic evolution followed by interrelations among functionally interconnected structures. This would be

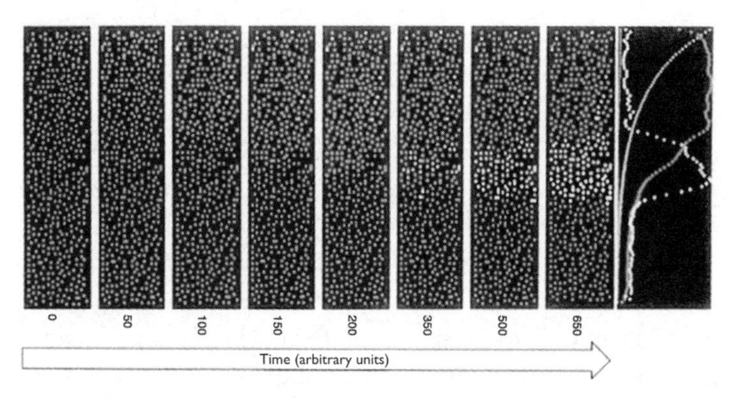

Figure 29: *Simulation of evolution over time in the expression of genes for the formation of borders or bands.* Evolution over time during embryonic development can also serve as a model for the evolution of the brain in, for example, the expansion of the frontal cortex in the mammalian brain. From M. Kerszberg and J.-P. Changeux, "A Model for Reading Morphogenetic Gradients: Autocatalysis and Competition at the Gene Level," *Proc. Natl. Acad. Sci USA* 91 (1994): 5823–5827; copyright © 2007 National Academy of Sciences, U.S.A.

valid for the expansion of the prefrontal cortex, from 3.5 percent in cats to 11 percent in macaques, 17 percent in chimpanzees, and 29 percent in humans, as well as for the differentiation of language areas, including the reorganization of the superior parietal cortex and growth of the frontal lobe in *Homo habilis* and hemispheric asymmetry in *Homo erectus*. These still hypothetical models would seem to merit more attention from molecular biologists.

CRITICISM OF THE NOTION OF A GENETIC PROGRAM

It remains difficult to resolve the evolutionary paradox of nonlinearity between genetic complexity and cerebral complexity, but one must realize that it is based on comparison of the genome of the fertilized egg with the organization of the adult brain. It is obvious that the complexity of the nervous system cannot be understood simply by study of the adult: it develops progressively during embryonic and postnatal life. Sydney Brenner, among other molecular biologists, emphasized the genetic program in the fertilized egg, but Gunther Stent justifiably questioned this. Obviously, the concept of a unitary program stems from cybernetics, in which models apply to bacteria, but in progressing from them to humans, one cannot just declare that organizational complexity corresponds to a lengthening of the program, as Jacob did. Not only does DNA possess a linear structure, whereas development is three-dimensional, but the term *program* suggests a single command center, which exists only as such in the fertilized egg. It "delocalizes" in the first stages of embryonic development.

At this stage it is useful to replace the schema of the organism as a cybernetic machine by that of the organism as a system (Ludwig von Bertalanffy, 1901–1972). The system, a group of elements interacting with each other and with the environment, is defined by enumeration of its constituent elements, the description of their states and transitions between these states, and the relationships between elements and their rules of interaction. The abstract notion of the magnetic tape program is thus replaced by an exhaustive description of an array of spatiotemporal processes. Passing from a theoretical vocabulary to one of observation, the element becomes the embryonic cell, the number, position, and state of which vary with time. The rules of interaction become reciprocal exchanges of signals between cells. The state becomes the catalogue of open or closed genes.

Abandoning the concept of a genetic program encourages a more precise and complete observation of reality, which emphasizes the contribution of cellular interactions in the development of the organism and the establishment of adult complexity. In such a context genetic determinism, if we preserve the expression, covers very different processes, whether it refers to the primary structure of a protein or to highly integrated faculties such as human language. In the first case there exists an unequivocal and unambiguous relationship between the sequence of bases in a structural gene and the sequence of amino acids in a protein. In the second we are dealing with a cerebral function that involves large cell groups assembled progressively over time and not necessarily synchronously: in this case we cannot attribute a gene to a function. So understanding the determinism of highly integrated functions like language means understanding that of communication during development, both between embryonic cells and then between neurons. The task is perhaps not as unrealistic as it may appear. Indeed, communication between cells could itself be determined by a small number of specialized genes: activity in some of these communication genes at critical stages of development could have major effects on the organization of a system. Jacob suggested that the simplicity of addition could be translated into the complexity of integration. The paradox of the complexity of the brain and a relatively modest change in the number of genes during human evolution thus becomes resolvable if the products of these genes affect integrated communication during cerebral development.

GENETIC MECHANISMS OF VARIATION OF THE BRAIN

In *An Introduction to Genetics* (1939), Alfred Sturtevant and George Beadle defined genetics as "the science of heredity and variation." An understanding of variation in the functional organization of the nervous system is thus necessary for the analysis of underlying genetic mechanisms. In a species as genetically heterogeneous as humans, there is significant individual variation in weight and even histology of the brain. Ray Guillery analyzed the latter in the lateral geniculate nucleus (LGN), the thalamic nucleus that relays information from the retina on its way to the visual cortex. He found that the arrangement of its layers is different in the albino, first in the cat in 1969 and then in a human case. Even in the normal human LGN, the number and organization of the layers is variable. The causes of such variation are unknown, except in the albino, where they are related to the known genetic mutation in this abnormality.

Individual variation due to genetic heterogeneity is considerably reduced or entirely absent among members of a clone, such as in the crustacean *Daphnia* or the fish *Poecilia,* or in a consanguineous line of *C. elegans,* or in the cricket or the mouse. In *Daphnia* and *C. elegans* the total number of neurons does not vary; exactly 258 have been counted in the latter. In vertebrates such as the mouse, variation in number of neurons is difficult to estimate in view of their very large number. Richard Wimer, Cynthia Wimer, and their colleagues in 1976, however, showed that there was variation in the hippocampus in inbred strains of mice, but that it was less than that observed between one consanguineous line and another. Variation in connectivity does not necessarily follow that of neuronal number. The number of synapses per neuron is commonly several thousand, and phenotypic variability in connectivity exists even in organisms with a fixed neuronal count. To estimate the details of connectivity, we need to study tissue with electron microscopy of serial sections of isogenic individuals, as described by Cyrus Levinthal and colleagues in 1976, and then reconstruct identifiable neurons and their processes. There is significant variation in the number and distribution of axonal branches and even the synapses they make.

Keir Pearson and Corey Goodman in 1979 managed to compare axonal morphology of a particular interneuron, the descending contralateral movement detector (DCMD) of the locust, by intracellular dye injection, and the functional state of its synaptic connections with the motor neurons controlling the leg muscles by electrical recording. They discovered that in most cases the axon of the DCMD stopped at the level of the first abdominal ganglion, whereas others reached the second or third. So there was phenotypic variation in functional connectivity among locusts of the same consanguineous colony. Even in species with a fixed number of neurons, and a fortiori in other cases, the functional organization of the nervous system is not exactly reproducible from one isogenic individual to another. This phenotypic variability in connectivity has been reported everywhere it has been looked for, and the hypothesis emerges that it represents a particular mode of development of the nervous system. We now turn our attention not to variability in a specific identifiable neuron in isogenic individuals but to variability in a single individual of given neurons in an apparently morphologically homogeneous population. Let us return to motor innervation in the locust. There are fast and slow extensor muscles in each leg. John Wilson and Graham Hoyle in 1978 compared the morphology of two particular motor neurons, A and B, found in a similar position in the three thoracic ganglia and with similar morphology. Whereas in the first two ganglia A innervated fast extensor muscles and B the slow ones, however, the inverse was found in the third ganglion. So apparently identical neurons derived from the same embryonic neuroblasts (the embryological precursors of the neurons themselves) acquire a different functional specialization.

Another example is the Purkinje cell in the cerebellum. In the adult cat there are several million Purkinje cells, and few morphological characteristics distinguish one from another. Nevertheless, in the region that receives the projection of the eye muscles, we find electrical responses that distinguish, for example, cells that respond to stretching of the right lateral rectus muscle (which moves the whole eyeball in an outward direction) by a reduced discharge, from those responding to stretching of the left medial rectus by increased discharge. Again there is functional specialization in an apparently morphologically homogeneous population. Each Purkinje cell possesses a singularity that allows it to be distinguished from its neighbors. This notion is particularly relevant if we trace the embryological origin of these cells, using for example the technique of chimeras. We find no direct relationship between the topology of Purkinje cells in the adult and the successive divisions of their embryonic precursors. During the migration that follows the final divisions of these precursors, the cells distribute themselves without obvious order in the Purkinje layer. When they reach their final position, they are equivalent, with the same state of differentiation and probably the same set of open genes that will diversify later. So singularization of a neuron consists of an increase in functional anisotropy, which distinguishes it from its neighbors. Singularization could result, for example, in differential localization of neurotransmitter receptor molecules, which are distributed uniformly over the surface of an embryonic cell but will later regroup beneath specific nerve terminals. When the first synaptic contacts form, they are much more numerous than they are in the adult. This synaptic exuberance is transitory, being followed by synaptic stabilization of certain privileged contacts that confer functional specificity, or singularity, on the neuron (see Figures 23, 24). One may anticipate that once the individual pattern formation becomes stabilized, a distinct pattern of genes is expressed (Tamas Bartfai, personal communication). In other words, the "singularity" of a given neuron within a class would manifest itself both at the connectivity level and in the pattern of genes expressed.

THE ROLE OF SPONTANEOUS ACTIVITY IN THE DEVELOPING NERVOUS SYSTEM

My hypothesis is that activity of a neuronal network during development intervenes in this singularization. I see this activity as the set of processes that directly or indirectly result in changes of electrical properties of a neuronal membrane. It includes propagation of an action potential, chemical synaptic transmission and modulation, and electrical coupling. There are several reasons for this choice. First, all electrical membrane phenomena can result in cellular integration. Next, in the context of a systemic model, activity in a developing neuronal network represents a form of interaction among elements, including cells, nerve centers, and sensory and effector organs, which ensures both integration and diversification. Through properties of convergence and divergence specific to a nerve network, this activity introduces a new combination of signals. In the end spontaneous activity that is present very early in embryonic development can be modulated or even replaced by evoked activity that results from interaction of the developing organism with the outside world. The array of endogenous signals is enriched by evoked signals, which can then participate in singularization, which in mammals and especially in humans continues long after birth, particularly in the cerebral cortex. The hypothesis of participation of nerve activity in regulation of embryonic development allows a considerable economy of genes in the elaboration of the nervous system.

Various electrical phenomena have been observed experimentally in the first stages of embryological development and even in the unfertilized egg. For example, the oocyte of the toad *Xenopus* responds to acetylcholine by depolarization and to dopamine and serotonin by hyperpolarization (Kiyoshi Kusano and colleagues 1977). In the axolotl, effective electrical coupling exists between the first blastomeres of the segmented egg. At the time of the formation of the neural fold (the invagination of the embryonic neural tissue that will form the nervous system), there is a difference in membrane potential between ectodermal cells (about –30 millivolts) and presumptive neurons (about –44 millivolts). When the neural tube detaches from the ectoderm to form the nervous system itself, their electrical coupling disappears. Regenerative activity, or rectification, appears in the membrane of neuroblasts, but there is no sign of an action potential at this stage (as described by Anne Warner in 1973).

Electrical activity of the embryonic nervous system has been studied mainly in the chick. In 1885 William Preyer noted that at three and a half days of incubation, the embryo displayed spontaneous movements. Initially unilateral, they progressively invaded the whole body and became alternating and periodic (type I). From the ninth to the sixteenth day widespread brief, arrhythmic, large-amplitude movements appeared: the embryo flexed and extended its legs, flapped its wings, opened and closed its beak (type II). From the seventeenth day until hatching, regular stereotyped coordinated movements developed: this type of behavior allowed the chick to break the shell and leave the egg (type III). The frequency of movements as a whole was greatest around the eleventh day, at twenty to twenty-five movements per minute: this was the time when motor innervation was being realized. In 1938 Rita Levi-Montalcini and Fabio Visintini found that these spontaneous movements were blocked by curare: they coincided with electrical activity of similar frequency in the spinal cord, leaving little doubt that embryonic motor activity is neural in origin. Electrophysiological recording on chick embryonic spinal cord on the fifth day demonstrated units with regular periodic discharges; others showed bursts. Large-amplitude, polyneuronal burst discharges with delayed activity were more and more frequent and were responsible for triggering movement, according to Robert Provine in 1972. Experiments involving early cervical and thoracic transection of the spinal cord at forty to fifty hours by Ron Oppenheim in 1973, as well as extirpation of sensory dorsal roots, showed that only type III movements are controlled supraspinally and require a reflex involving three synapses. Type I and II movements are strictly spinal in origin and result from spontaneous electrical activity of motor neurons.

This hypothesis of a possible role of spontaneous electrical activity in neuronal singularization makes sense only if it does not cost more in structural information than the activity it is supposed to determine. This is why I sought to define the minimum molecular elements responsible for spontaneous electrical activity. I was helped in this task by the observation that this activity is usually oscillatory. First, the thermodynamic conditions for the appearance of oscillations were defined. For Ilya Prigogine and his colleagues the oscillations could take place only in an open thermodynamic system that exchanged energy continuously with the outside world. Further, this system had to be in a stable system, but not in equilibrium. These dissipative structures could appear when nonlinear relations existed between flux and forces that were due, for example, to the existence of cooperative interactions and effects of feedback between elements of the system. We must distinguish two types of oscillation in the nervous system. Some appear in neuronal chains and depend directly on their organization, such as circuits controlling respiratory movements in vertebrates or walking in certain insects. Others appear in isolated cells, such as the oscillations responsible for spontaneous embryonic activity. Among these cellular oscillations we distinguish cytoplasmic from membrane oscillators.

The enzymatic chain of glycolysis is an example of a cytoplasmic oscillator. It produces sustained oscillations both in vivo, as in yeast, and in vitro, as in an acellular extract of muscle (Benno Hess, 1979; Albert Goldbeter, 1980). Phosphofructokinase plays a critical role in these oscillations, as it is the object of negative feedback by adenosine triphosphate (ATP), the final product of the chain, and positive feedback both by its substrate fructose-6-phosphate (cooperative effect) and by the low-energy precursors of ATP: adenosine diphosphate (ADP) and adenosine monophosphate (AMP). The enzyme oscillates between two extreme states: in the activated state the substrate (positive effector) disappears and the level of ATP (negative effector) increases until the enzyme finds itself in the inhibited state. In the inhibited state the substrate and the ADP (positive effectors) accumulate until the reactivation of the enzyme. The regulatory properties of the enzyme confer the nonlinearity required for the appearance of oscillations in a system that finds itself open because of the permanent inflow of substrate.

The bursting neurons of the mollusk, like the R15 cell of the abdominal ganglion of *Aplysia* (Felix Strumwasser, 1965; Robert Meech, 1979), constitute one of the best-known examples of the membrane oscillator. The R15 cell, even isolated from its afferents, produces bursts of ten to twenty action potentials every five to ten seconds, like clockwork. These bursts are superimposed on a base oscillator composed of two slow channels, which are distinct from those involved in an action potential. One is selective for calcium, and its opening is sensitive to potential and is accompanied by entry of a depolarizing calcium or sodium current. The other is selective for potassium, and its opening is activated by the calcium present inside the cell and is accompanied by the exit of a hyperpolarizing potassium current. The regenerative depolarization (positive cooperative effect) associated with the opening of the slow calcium channel causes increased intracellular calcium concentration, which, by activating the slow potassium channel (positive feedback), causes an opposite change in potential to the one that permitted its entry. These regulatory cycles lead to slow oscillations. When the membrane potential passes the threshold for an action potential, bursts of discharges appear at the crests of the slow waves. These discharges open fast ion channels, which causes ionic changes that modulate the base oscillator. Opening these four types of channels leads to passive loss of ions that are compensated for constantly by intense active pumping by membrane ATPases. This very simple membrane system can be categorized as an open thermodynamic system.

The schema described with the base oscillator of mollusk burst neurons can be applied with little modification to oscillations of potential of the Purkinje fibers in the heart or secretory ß cells of the pancreas. It probably accounts also for the spontaneous bursts in the embryonic spinal cord. The genetic cost for spontaneous activity of this type is clearly no more than a few structural genes.

SPONTANEOUS FETAL ELECTRICAL ACTIVITY AND PARADOXICAL SLEEP

Spontaneous electrical activity has always been found in vertebrate embryos and fetuses whenever it has been sought (Figure 30). In humans especially, spontaneous movements appear during the second month, when the embryo is still only four centimeters long. They involve the head and trunk and resemble type I movements in the chick embryo. During the third and fourth months the movements become more generalized and the first reflexes occur, together with facial expressions and sucking and swallowing movements: the fetus drinks amniotic fluid. The movements are perceived by the mother and have been shown to be spinobulbar (involving the spinal cord and brainstem) in origin. They present analogies with chick type II activity. From the fourth to the sixth month more-specialized activities develop, such as regular and stable respiratory movements. They have a suprabulbar component (that is, involving the cerebral hemispheres, including the cortex) and are analogous to type III activity in the chick. As occurs in the chick, a phase of acceleration, then deceleration, takes place during fetal development; the fetus makes two to a maximum of ten movements per minute around the eighth month. Of course, spontaneous electrical activity accompanies these fetal movements. The first recordings were obtained from the brainstem of the seventy-day fetus by Bergstrom in 1969. The activity continues and diversifies during subsequent months (Figure 30).

From a strictly methodological point of view we must

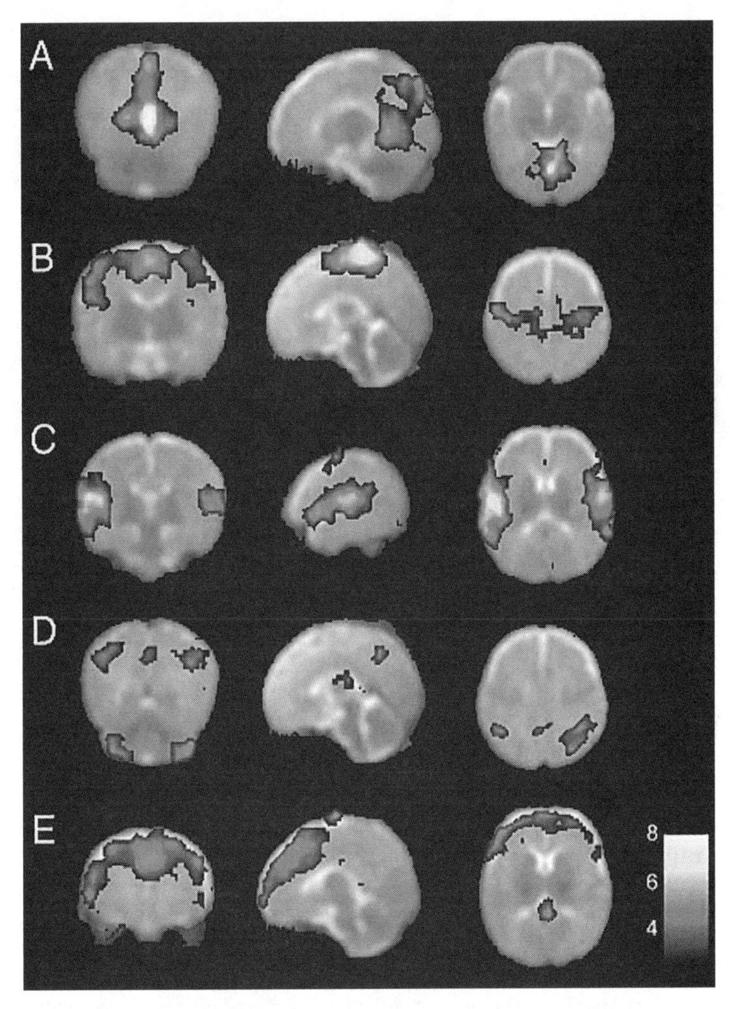

Figure 30: *Spontaneous activity in the brain of a neonate.* At twenty-five weeks of gestation, using fMRI, one can distinguish five principal circuits, one of which (E) resembles the conscious neuronal workspace circuit. From P. Fransson et al., "Resting-State Networks in the Infant Brain," *Proc. Natl. Acad. Sci. USA* 104 (2007): 15531–15536; copyright © 2007 National Academy of Sciences, U.S.A.

note that such research was undertaken in totally different conceptual frameworks, depending on whether chicks or human fetuses were the subjects. The human electroencephalograph returns a rather global appreciation of cerebral electrical activity—and, therefore, the more obvious states such as waking and sleeping. Studies of sleep in the adult have revealed command centers with periodic oscillating activity consisting of groups of neurons in various nuclei of the pons region of the brainstem. By their richly divergent axonal ramifications, they distribute signals to a large number of regions, even the whole brain. The locus coeruleus controls waking, and the anterior raphe regulates slow-wave sleep. Jouvet's paradoxical (or REM) sleep involves high-voltage spontaneous electrical activity resulting in eye movements but a generalized drop in muscle tone, so that, except in the embryo, this sub-coerulean activity does not result in movements except of the eyes. They can occur only after destruction of a medullary center that blocks them. The sub-coerulean oscillator becomes active late, in the cat between ten and twenty days after birth (Edgar Adrian, 1978). It replaces in a way the medullary oscillators of the embryo and fetus. It is possible, but not proved, that paradoxical sleep, perhaps indicating dreaming, plays a role in the adult similar to that of spontaneous activity in the embryo—for example, to maintain neuronal singularity disturbed by diurnal activity, or on the contrary to consolidate a track created by interaction with the environment.

In the embryo nerve activity is essentially endogenous. As sense organs become functional, evoked responses replace spontaneous activity, at least during waking. In humans touch develops early. From day 49 of gestation the fetus responds to tactile stimulation of the lips, and between the fifth and seventh months sensory innervation of the fingers is practically complete. The vestibular apparatus becomes functional between 90 and 120 days, and auditory sensation, as judged by changes in heart rate, between 180 and 210 days. A flash of light evokes potentials in the premature infant at 29 weeks. These fetal sensory functions are far less efficient than those in the adult. Maturation of sense organs, characterized by a reduced threshold of sensitivity and decreased latency, continues long after birth. At birth the human brain weighs about 300 grams. It increases about fivefold to reach its adult weight at some seven years of age. By comparison, the weight of the chimpanzee brain at birth is already about 60 percent of its adult weight. In the cerebellum and cerebral cortex, most dendritic arborization and synaptogenesis are postnatal. In 1975 Brian Cragg estimated that in the primary visual cortex of the cat, the average number of synapses per neuron increased from a few hundred to 12,000 between day 10 and day 35 after birth. Then this number decreases between 20 and 30 percent to reach its adult value. It is thus legitimate to think that activity evoked by interaction with the environment replaces spontaneous activity and thenceforth regulates synaptic development.

CRITICISM OF THE EMPIRICIST POSITION

One might think from an empiricist point of view that interaction with the outside world provokes activity in the nervous system, the characteristics of which are directly related to the types of physical signal received by the sense organs. In fact, this is not so. Action potentials in the auditory nerve are identical in nature to those in the optic nerve. Cellular analysis of the function of sense organs allows us to go further. In most cases the effect of the physical signal causes modulation of spontaneous activity, which obviously predates interaction with the environment. For example, in 1968 Horace Barlow showed that in the cat, illumination of the retina causes an increase in the activity of ON ganglion cells from forty to seventy impulses per second. The first signs of spontaneous activity appear eight days after birth, that is, before any response to light. Another example is in the vestibular nerve, where there is considerable spontaneous activity in the adult cat: when the position of the head changes, this activity varies from five to thirty-five impulses per second. Ian Curthoys in 1979 showed that in the rat there is spontaneous activity in the vestibular nerve from birth, and it increases almost six times during the first twenty postnatal days. The sensitivity of the signal evolves in parallel. One can record signs of this activity in the neurons of the vestibular nuclei and even in the cerebellar cortex. So sensory receptors constitute peripheral oscillators, the activity of which contributes largely to the spontaneous activity recorded in the nervous system: interaction with the environment modulates the rhythm of peripheral sensory oscillators, which then regulate central oscillators.

One can imagine the molecular mechanism by which evoked activity in peripheral oscillators could modulate activity of central oscillators, more or less for the long term. The provision of this "track" in neurons would require only a small number of macromolecules, receptors, and cyclases, which in addition could be used for different tasks in other tissues. The cost in structural genes would also be limited. Contrary to certain biochemical theories of memory, no synthesis of new types of molecule would be necessary to produce lasting modification of neuronal properties following interaction of an organism with its environment.

V
Molecules and the Mind

The Discovery of Neurotransmitter Receptors

Neurotransmitters and their receptors appeared very early in animal evolution, and their structural genes have not changed much since. We have even found ancestors of the nicotinic acetylcholine receptor in ancient bacteria! Our understanding of this receptor has progressed greatly in the last few years, and it remains easily the best known of all neurotransmitter receptors and is one of the best reference models.

Since ancient times we have accepted the concept of active chemical substances in the human body. In 1877 Emil du Bois-Reymond suggested two possible mechanisms: "At the boundary of the contractile substance, either there exists a stimulatory secretion . . . or the phenomenon is electrical." In 1905 John Langley, revisiting the experiments of Claude Bernard, studied the comparative effects of nicotine and curare on muscle; the first, an agonist, caused muscle contraction, and the second, an antagonist, blocked the effect of this agonist. He described a localized, sensitive area immediately below the nerve terminals, even after denervation. He concluded that "the muscle substance which combines with nicotine or curari is not identical with the substance which contracts. It is convenient to have a term for the specially excitable constituent, and I have called it the receptive substance." So, from one tissue to another, differences in receptive characteristics would be due to different receptors. That same year Langley's student Thomas Elliott stated that "the effect of adrenalin upon plain muscle is the same as the effect of exciting the sympathetic nerves supplying that particular tissue." Many scientists then accepted the idea that a chemical agent was released at a nerve terminal. The concepts of a pharmacological agent and a receptor took on even more significance when linked to a third, distinct concept, that of chemical neurotransmission. Thus evolved a concept of a neurotransmitter stimulating a receptor in the muscle membrane opposite the nerve ending (Figures 31, 32). Several decades were necessary, however, for the concept of chemical synaptic transmission to be accepted definitively, even if it might be in parallel with an electrical form of transmission.

The evidence for acetylcholine as a neurotransmitter at the neuromuscular junction can be summarized as follows. Acetylcholine is the most active natural or synthetic substance in terms of muscle contraction, according to Henry Dale in 1914. Then in 1926 Otto Loewi and Ernst Navratil showed that stimulation of the vagus nerve slows the heart when it secretes a chemical that is degraded by acetylcholinesterase; this degradation is blocked by eserine. Finally, Henry Dale and his colleagues in 1936 showed that acetylcholine is liberated by the nerve terminal, and repeated stimulation of the nerve causes depletion of acetylcholine in the terminal when its synthesis is blocked by hemicholinium.

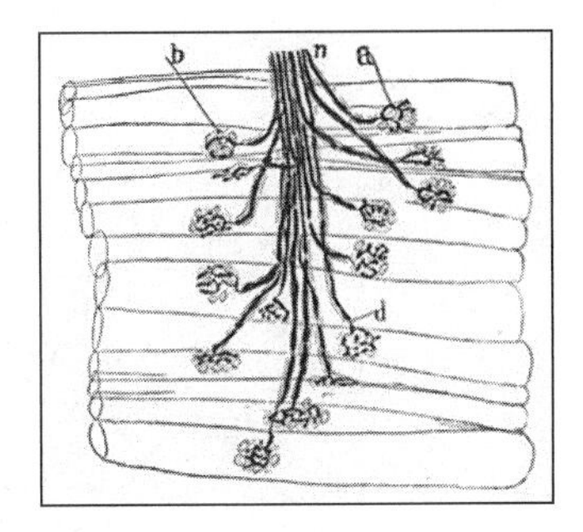

Figure 31: *The motor end-plate.* One can see a synapse between a motor nerve and skeletal muscle. Each nerve fiber makes a single synapse with a muscle fiber in the adult. From S. Ramón y Cajal, *Histologie du système nerveux de l'homme et des vertébrés* (Paris: Maloine, 1909), 1:372.

The Contribution of Electrophysiology

The first research following the elaboration of the concept of a receptor concerned the quantitative analysis of the response to a given pharmacological agent. An isolated organ, such as an intestine or a skeletal muscle, was connected to a recording apparatus, typically a smoked drum. The kinetics and amplitude of the response to various dilutions of active substance were measured. The transmitter was not changed by its reversible binding to the receptor, which was enough to trigger the response. The effect of agonists could be blocked by structural analogues without activating the receptor but decreasing the affinity for an agonist. They were referred to as competitive antagonists and supposedly bound to the same site as the agonist in a mutually exclusive fashion without causing a response. Other blocking agents decreased the amplitude of the response without modifying the affinity: they were noncompetitive and supposedly bound to sites other than the active site. Analysis of such experiments encountered two main difficulties: the measured response was only indirectly related to the initial agonist-receptor reaction and not directly proportional, and the kinetics were limited by diffusion of active substance. Electrophysiological recording, developed from the 1930s, allowed

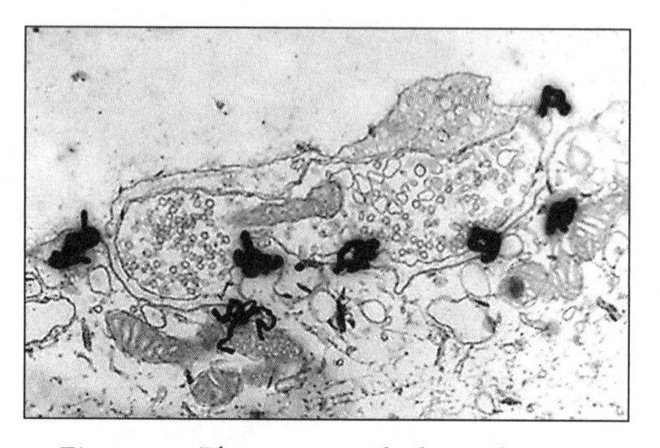

Figure 32: *The motor end-plate.* Electron microscopy shows a synaptic cleft between the nerve terminal (containing vesicles filled with acetylcholine) and the postsynaptic membrane, where the acetylcholine receptor is situated. Here the acetylcholine receptor in the electric eel is labeled with radioactive snake venom (black grains). From J.-P. Bourgeois et al., "Localization of the Cholinergic Receptor Protein in *Electrophorus electroplax* by High Resolution Autoradiography," *FEBS Lett.* 15 (1972): 127; reprinted with permission from Elsevier.

researchers to overcome this double difficulty. In 1938 H. Göpfert and H. Schaefer used extracellular electrodes to record the first indirect response of a muscle fiber to stimulation of its nerve, the postsynaptic, or end-plate, potential. This was a depolarization of 10 to 20 millivolts, which rose rapidly after a synaptic delay of about 0.3 milliseconds and declined to half in 1 to 3 milliseconds. We can simulate this effect by injecting acetylcholine iontophoretically into the muscle end-plate with a fine pipette. Thanks to this method, José del Castillo and Bernard Katz in the 1950s analyzed in detail the action of various cholinergic agonists and antagonists and proposed a kinetic model directly inspired by biochemical work on acetylcholinesterase:

$$R + A \leftrightarrow RA \leftrightarrow RA^* \text{ (open)}$$

where R is the receptor in the strict sense, A the acetylcholine substrate, and RA* the depolarizing component.

In addition to considerable improvement in resolution over time, electrophysiology provided essential information on the nature of the short circuit (Paul Fatt and Bernard Katz, 1950) provoked by acetylcholine, which, at the vertebrate endplate, causes selective increase of cation conductance for sodium, potassium, and calcium through what was called an ion channel. Changes to the chloride anion conductance occur in the inhibitory receptor for gamma aminobutyric acid (GABA). Taking a measure of the maximum membrane conductance as an index of the number of ion channels opened by the agonist, we can now construct dose-response curves that are much closer to the basic mechanism of the response than those obtained by standard pharmacological methods. Apparent dissociation constants obtained are 20 to 40 μM for acetylcholine. This order of concentration is close to that of acetylcholine in the synaptic cleft during the passage of a signal (0.3 to 1 mM) (Stephen Kuffler and Doju Yoshikami, 1975; Katz and Ricardo Miledi, 1977).

Nevertheless, one of the insuperable limitations of these electrophysiological recordings is that they refer exclusively to the only ionic response susceptible to causing modification of the electrical parameters of the membrane. The inference of binding data of the neurotransmitter from the response it produces can therefore be suggested only as a hypothesis. The elucidation of the mechanism of the opening of the channel by acetylcholine necessarily requires the direct measurement of its binding, which means using biochemical and physico-chemical methods on a system that can henceforth be entirely in vitro. Further, it remains necessary to imagine a molecular mechanism capable of explaining the transduction of the chemical signal into an electrical signal at the elementary level. This could be inspired by research carried out in a completely different domain, that of the molecular biology of bacterial regulation enzymes, which are subject to regulation not by a neurotransmitter but by an intracellular metabolic signal.

Is the Acetylcholine Receptor an Allosteric Protein?

In 1961 my work on the bacterial regulatory enzymes L-threonine deaminase, and that of John Gerhart and Arthur Pardee in 1962 on aspartate transcarbamylase, resulted in three sets of observations.

> *The inhibition of activity of these enzymes* by the end product of the metabolic chain acted as a regulatory signal by negative feedback. The antagonism between the substrate that binds to the active site and this end product is apparently competitive, in spite of the fact that these two components possess different steric structures.
>
> *The presence of cooperative effects* for binding of the substrate and/or the regulatory effector, as was found for the binding of oxygen to hemoglobin.
>
> *Abolition by chemical or physical methods* of the sensitivity to the regulatory signal without loss of enzymatic activity, but with concomitant uncoupling of cooperative effects, wrongly called "desensitization" at the time.

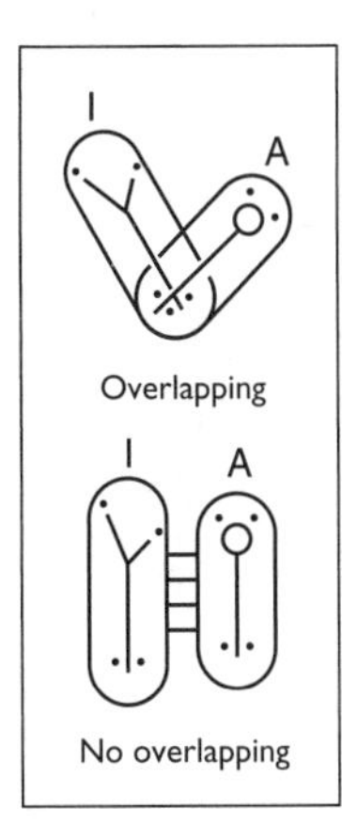

Figure 33: *Allosteric sites and transitions.* The first model from 1961 illustrates elementary regulatory interaction between the substrate threonine (A) and the regulatory inhibitor of threonine deaminase in *E. coli,* isoleucine (I). Overlapping indicates interaction by steric hindrance. No overlapping indicates allosteric interaction between topographically distinct sites. From J.-P. Changeux, "The Feedback Control Mechanisms of Biosynthetic L-Threonine Deaminase by L-Isoleucine," *Cold Spring Harbor Symp. Quant. Biol.* 26 (1961): 313–318; reprinted with permission from Cold Spring Harbor Laboratory Press.

These data were interpreted according to a model distinct from the classic model of direct interaction by steric hindrance. The latter model postulated that two categories of topographically distinct sites intervened in specific binding of substrate and regulatory signal, and that their interaction was indirect or allosteric (Jacob and Monod, 1962; Figure 33). A conformational, or allosteric, transition ensured coupling between different sites. In 1963 Monod, Jacob, and I envisaged this transition as resulting from an induced fit, caused by the binding of the regulating effector and/or the substrate (Daniel Koshland, 1959), which was manifest by a change of aggregational state of the protein molecule.

In 1965 Monod, Jeffries Wyman, and I adopted a radically different point of view (Figure 34). Allosteric transitions, instead of being induced by binding of ligands, were supposed to preexist binding in the form of a conformational equilibrium R ↔ T between an active relaxed state (R) and an inactive constrained or tense one (T). The protein molecule did not change aggregation state with the transition. It was supposed to be composed of identical subunits organized as

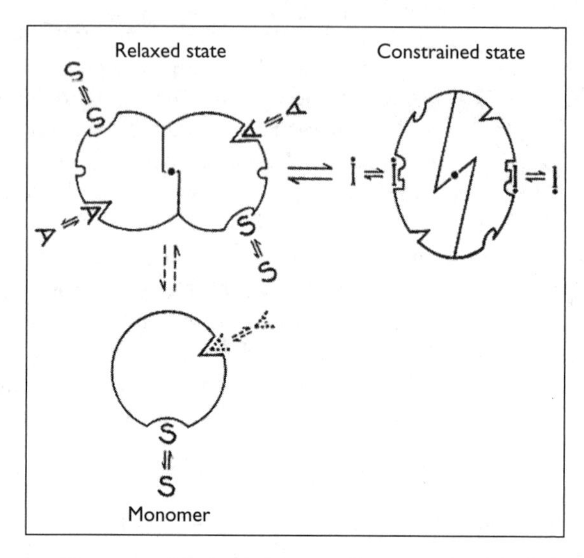

Figure 34: *Monod-Wyman-Changeux model of conformational transition.* From J.-P. Changeux, "Sur les propriétés allostériques de la L-thréonine désaminase de biosynthèse," *Bull Soc. Chim. Biol.* 47 (1965): 115–139 (copyright © The Nobel Foundation 1965).

oligomers possessing at least one axis of symmetry, and the conformational transition was supposed to alter the relationships between subunits (quaternary constraint) while conserving the symmetrical properties of the molecule. The affinity of one or several ligands changed at the transition from one state to another. In other words, signal transduction was mediated by displacement from a preexisting conformational equilibrium to the state for which the ligand had the greatest affinity. In consequence, one could distinguish a function of state R and binding T that expressed the fraction of protein in state R and the fraction of sites occupied by the ligand. This property distinguished unambiguously this concerted selective, or Darwinian, model from other sequential, or Lamarckian, models on the basis of the induction of conformational change by the ligand, such as that described by Koshland, Nemethy, and Filmer in 1966. Structural studies undertaken since the proposal our 1965 model have in general confirmed its validity, with certain exceptions noted by Max Perutz in 1989.

Neither our 1963 nor our 1965 model alluded to the possibility that chemico-electrical transduction at the synapse could bring into play allosteric mechanisms at the level of excitable membranes. I first mentioned this idea in 1964 and developed it later. It is accompanied by two propositions:

> *The transduction of chemical to electrical signal* involves a transmembrane protein that contains a receptor domain that binds the transmitter and a biologically active domain consisting of an ion channel. Their coupling is assured by a conformational transition of the molecule (see Figure 34).
>
> *The cooperative effects observed in the response to the transmitter* result either from the organization of an unlimited network of receptor elements or from an association in transmembrane oligomers with an axis of symmetry perpendicular to the membrane (Figure 35).

The experimental testing of the proposals necessitated the identification in vitro of a receptor engaged in electrochemical transduction. The first to be isolated was the acetylcholine receptor, which was also the target of the widely consumed drug nicotine, hence the name nicotinic receptor.

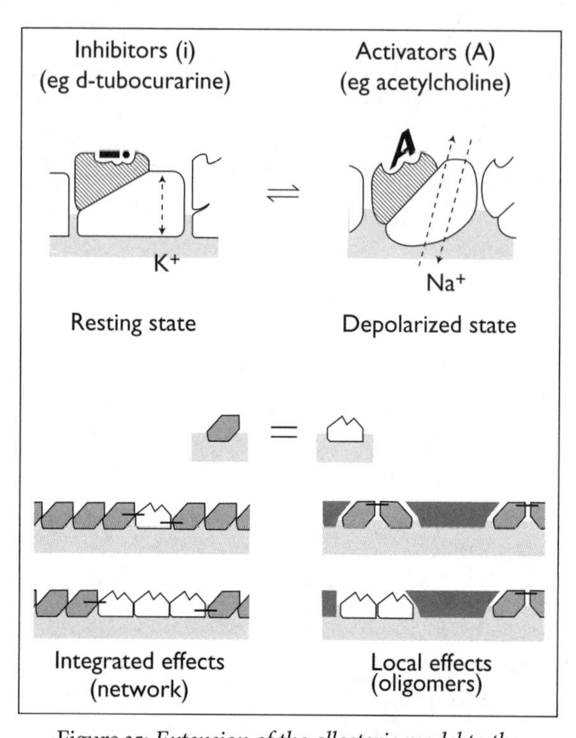

Figure 35: *Extension of the allosteric model to the acetylcholine receptor.* We propose that the excitatory and inhibitory binding site is allosteric, and that the active site is the channel through which sodium and potassium ions pass. The receptor can exist in a resting state with the channel closed, and in a depolarized state with the channel open. The integration of the receptor in the membrane imposes structural constraints: a network of receptor elements or association in transmembrane oligomers with an axis of symmetry perpendicular to the membrane. From J.-P. Changeux, "Allosteric Model Figures: Membrane Allosteric Receptors, Extension of Allosteric Model to the Acetylcholine Receptor," in *Symmetry and Function of Biological Systems at the Macromolecular Level: Proceedings of the 11th Nobel Symposium Held August 26-29, 1968, at Sōdergarn, Lidingō, in the County of Stockholm,* ed. Arne Engstrom and Bror Strandberg (New York: John Wiley and Sons, 1969); copyright © The Nobel Foundation 1969.

The Electric Organ of *Torpedo*: Identification of the Acetylcholine Receptor

In 1970 our first evidence of direct binding to the acetylcholine receptor was made possible by employing a very favorable tissue, the electric organ of the electric fish or *Torpedo*. It is specialized to give electric shocks (250 volts at 0.5 amps for the electric fish, 20 to 60 volts at 50 amps for *Torpedo*) and consists of multinucleated cells, the electroplaques or electrocytes, each receiving hundreds or thousands of identical synapses, a total of some 10^{11} to 10^{12} synapses per organ. This electrical tissue is both extremely rich in cholinergic synapses and homogeneous, which renders it particularly useful for biochemical studies, as already described by David Nachmansohn in 1936. Another advantage is that the electroplaque is a giant cell that can be dissected, and its electrical responses to active pharmacological agents can easily be recorded. Thus, the electroplaque represents a missing link between physiology and biochemistry. It is also suitable for subcellular fractionation. Successive centrifugation of homogenates enabled us to purify membrane fragments, which closed up to form vesicles or microsacs. When equilibrated with labeled ions, these microsacs responded to the presence of acetylcholine by increased passive ionic flow. The pharmacology of this in vitro response was identical to that of the electroplaque. Such membrane fragments could thus serve to identify the receptor site by binding techniques, and also to isolate the molecule that carries it. The first attempts to label the receptor site were hindered by the multiplicity of binding sites of the radioactive ligands used and by a lack of specificity. A decisive stage in the identification of the site was reached with the use of a different type of label, small toxic proteins of 6,000 to 7,000 daltons that are present in the venom of certain snakes, such as *Bungarus* (Chen-Yuan Lee and C. C. Chang, 1966). One of them, α-bungarotoxin, fixes very selectively and with high affinity to the receptor site. It serves to identify the site either by displacing the radioactive cholinergic agonist decamethonium (in our studies; Changeux, Kasai, and Lee, 1970), or by direct binding after radioactive labeling (Miledi et al., 1971).

The molecule carrying this site was solidly bound to the membrane and could be extracted in an aqueous solution using mild detergents without losing its binding properties. It could then be purified to homogeneity by affinity chromatography on beads of α-bungarotoxin or immobilized cholinergic ligands. Four facts showed unambiguously that the purified protein was indeed the physiological receptor.

> *Cholinergic agonists and antagonists* were bound by this protein with the same order and range of affinity as recorded by electrophysiological techniques with the isolated electroplaque.
>
> *Immunization of an animal with fish receptor* triggered an autoimmune paralysis analogous to human myasthenia gravis, and the resulting antibodies blocked the physiological response (Jim Patrick and Jon Lindstrom, 1973).
>
> *Antireceptor antibodies and α-bungarotoxin* selectively labeled the receptor protein at the postsynaptic membrane.
>
> *Purified protein reinserted* in artificial lipidic microsacs regulated ionic flux with the same characteristics as when it occurred in the postsynaptic membrane in vivo. Thus, the purified protein contained the ion channel and the receptor site of acetylcho-

line, as well as the allosteric coupling mechanism between these two sites.

Purified receptor protein obtained from the electric organ has a molecular weight of 290,000 daltons and results from assembly of five subunits as a pentametric oligomer (Hucho and Changeux, 1973). This $\alpha_2ß\gamma\delta$ oligomer is composed of four chains of apparent molecular weight 40,000 (α), 50,000 (ß), 60,000 (γ), and 66,000 (δ) daltons (J. A. Reynolds and Arthur Karlin, 1978; Michael Raftery et al., 1980). It contains two copies of acetylcholine and α-bungarotoxin binding sites. The α chains are labeled by covalent affinity ligands of the receptor site and therefore contain all or part of this site. Noncompetitive blockers of the ionic response, supposed to bind at the level of the ion channel itself, bind to a site common to the four chains of the receptor and are present as a single copy per oligomer. The four chains of the receptor cross the membrane and form compact, roughly cylindrical bundles. Seen head-on by electron microscopy (Jean Cartaud et al., 1973; Robert Stroud et al., 1985), the light form is a centered rosette nine nanometers in diameter, at the level of which appear five unequal masses corresponding to the five chains, of which the two smallest, which are nonadjacent, are reinforced by α toxin and can therefore be identified as α chains. Studies by Nigel Unwin in 2000 using high-resolution electron microscopy and X-ray crystallography provided a detailed image of the structure of the receptor at the atomic level (Figures 36, 37).

Molecular Genetics of the Acetylcholine Receptor

The elaboration of a paradigm for specific binding at the receptor site and the purification of the receptor protein were decisive steps in research on the acetylcholine receptor. A

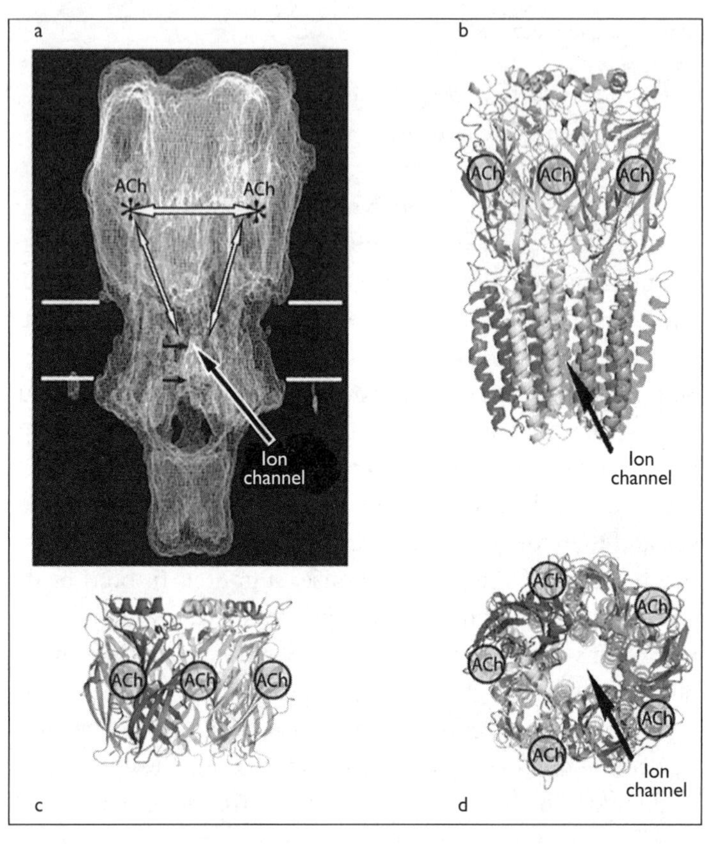

Figure 36: *The acetylcholine receptor.* (a) High-resolution microscopy of the acetylcholine receptor molecule. The binding sites for ACh and the ion channel are topographically some distance apart, and so their interaction is allosteric. From N. Unwin, "Nicotinic Acetylcholine Receptor and the Structural Basis of Fast Synaptic Transmission," *Philosophical Transactions of the Royal Society of London, Biological Sciences* 355 (2000): 1813–1829; reproduced with permission. (b) Molecular model of the neuronal α7 acetylcholine receptor, with perfect pentameric symmetry. (c, d) Molecular model of the acetylcholine receptor. Note the localization of the binding site at the interface between two subunits. From J.-P. Changeux and A. Taly, "Nicotinic Receptors, Allosteric Proteins and Medicine," *Trends in Molecular Medicine* 14 (2008): 93–102; reprinted with permission from Elsevier.

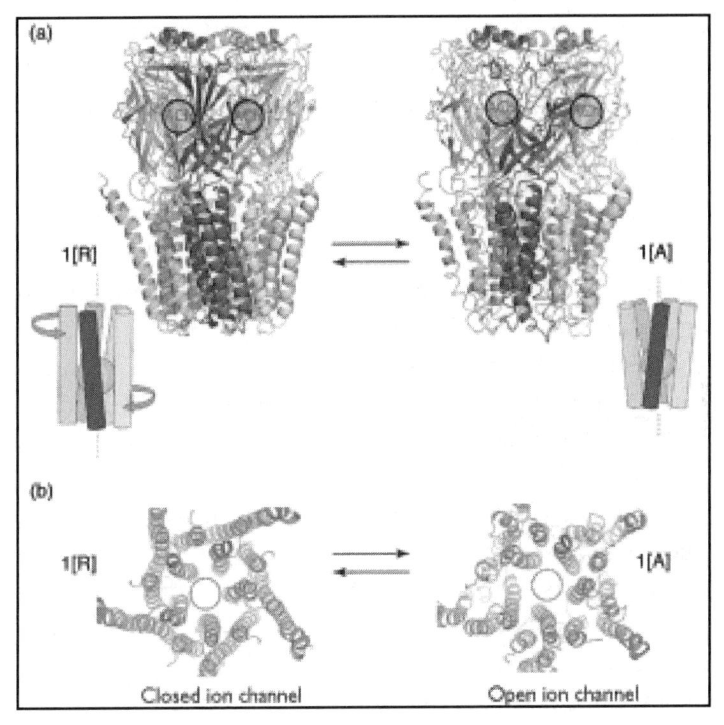

Figure 37: *Model of allosteric transition by quaternary torsion.* Resting and active states (R, A) are shown in lateral views (a) with movement of subunits. The perpendicular views (b) show open and closed ion channels. From J.-P. Changeux and A. Taly, "Nicotinic Receptors, Allosteric Proteins and Medicine," *Trends in Molecular Medicine* 14 (2008): 93–102, reprinted with permission from Elsevier.

third equally fundamental step was the application of molecular genetics to these studies. Purification on a large scale enabled us in 1979 to obtain sufficient quantities of the α subunit of the *Torpedo* receptor for the sequence of NH_2-terminal amino acids to be determined by the automated degradation method of Pehr Edman. Remarkable sequence homologies, from 35 to 50 percent identical, were subsequently reported between the four chains of the receptor by Raftery and his colleagues in 1981. Once these data were acquired, complementary DNAs (cDNA) to the cytoplasmic messenger RNA were cloned and the protein sequence of each subunit determined in 1982 by the teams of Masaharu Noda and his colleagues in Kyoto, by Stephen Heinemann in San Diego, and by ourselves at the Pasteur Institute. The sequences present notable homologies over their whole length, in accord with the hypothesis of evolution from a common ancestral gene suggested by Raftery. The receptor pentamer is a pseudosymmetrical oligomer with a fivefold rotational axis, as one might expect with an allosteric protein.

Analysis of the distribution of amino acids along the sequence of the four chains leads to subdivision into several domains: two hydrophilic, one large and one small, and four hydrophobic. The large hydrophilic NH_2-terminal domain is on the outer, synaptic side of the membrane, the small one on the inner, cytoplasmic side. The four hydrophobic domains, each with about twenty amino acids, form transmembrane α-helixes. The ion channel results from assembly of the five chains of the receptor at the level of their common part in the axis of pseudosymmetry of the molecule. The expression of a functional receptor from cloned cDNA coding for each of the receptor chains was obtained by Masayoshi Mishina and his colleagues in 1984. When the messenger RNAs of the four chains are introduced into a *Xenopus* oocyte, one obtains a receptor that binds α-bungarotoxin and produces ion channels sensitive to acetylcholine in a way similar to the native receptor's. The 290,000-dalton receptor protein thus suffices to cover the principal biophysical and pharmacological properties of the physiological response to acetylcholine.

Functional Properties of the Acetylcholine Receptor

In parallel with acquisition of these structural data, which are now at the atomic level, our knowledge has advanced concerning the functional properties of the receptor. For a long time these data were based exclusively on measuring electrical parameters, the limitations of which I have already outlined. Important progress in this domain was, however, realized with the remarkable increase in resolution in analysis of the physiological effect of acetylcholine. In 1970 Katz and Miledi wrote that it seemed possible that during the steady application of acetylcholine to the motor end-plate, "the statistical effects of molecular bombardment might be discernable as an increase in membrane noise, superimposed on the maintained average depolarisation." This noise, directly associated with the effect of agonists, was recorded and analyzed in terms either of discharge effects (Katz and Miledi, 1973) or of all-or-nothing opening and closing of discrete channels (C. R. Anderson and Charles Stevens, 1973). Recording of fluctuations in single channels that had the form predicted by this second model was later achieved by Ervin Neher and Bert Sakmann in 1976. It then appeared that from one agonist to another of the muscle receptor, the amplitude of the basic fluctuation, of conductance γ, did not vary, whereas the mean time of opening changed.

Opening of ion channels, or activation, takes place in physiological conditions when a chemical impulse of highly concentrated acetylcholine is liberated very briefly. If the transmitter is applied to the postsynaptic membrane for several seconds or minutes at weak concentrations, the amplitude of the permeability response diminishes reversibly as a function of time: there is desensitization as described by Katz and

Stephen Thesleff in 1957. The processes of activation and desensitization have been reproduced in vitro by measuring ionic flux in preparations of purified membranes and even in a functionally reconstituted purified receptor in lipid bilayers, by ourselves and Robert Anholt and his colleagues in 1981. The curve of variation of initial flux with the concentration of acetylcholine is slightly sigmoid, with an apparent dissociation constant of 40 to 80 μM; it practically overlies the dose-response curve obtained from electrophysiological measurements using voltage clamping. In the same way, the two phases of desensitization, fast and slow, can be seen in vitro. Fragments of purified membrane and reconstituted receptor therefore possess the same functional properties of the receptor-channel complex that are found in vivo. These systems have the advantage of permitting the measurement of the binding of agonists and the relationship of this binding with the opening of the ion channel.

Early studies of equilibrium binding of acetylcholine and other cholinergic ligands with a *Torpedo* receptor revealed a very high affinity of acetylcholine for the receptor site at concentration levels three orders of magnitude lower than the concentration of acetylcholine in the synaptic cleft during impulse transmission. Such high affinity is incompatible with repeated stimulation of the motor end-plate. In reality, our rapid-mixing kinetic analysis of binding of acetylcholine (Michel Weber et al., 1975), and of fluorescent analogues to receptor-rich membranes (Thierry Heidmann and Changeux, 1979), showed that at rest and in the absence of an agonist, the receptor binds acetylcholine with low affinity and that fast mixing with acetylcholine triggers a cascade of transitions that slowly result in a state of high affinity of the receptor. A four-stage model accounts for such binding kinetics (Figure 38). R is the resting state, A is active, I is intermediate, and D is desensitized. Affinity for acetylcholine increases from R to A to I to D, where R to A is very fast, whereas transitions to I and D are, respectively, in time scales from fractions of a second to several seconds or minutes. In 1983 we compared these binding data with ionic flux data at the same time scale and suggested that in state A the channel is open (active state) and that it is closed in states of high-affinity I and D (desensitized states). Following the Katz-Thesleff model of 1957, desensitization of the response is identifiable with slow transition toward the inactive, high-affinity state. In addition, this schema agrees with the model of concerted transition proposed for allosteric proteins by Monod, Wyman, and Changeux in 1965. In particular, state D exists spontaneously before agonist binding (20 percent at rest) and can be stabilized at equilibrium by allosteric ligands other than the agonist—for example, noncompetitive blockers. Kinetic analysis of covalent attachment of a noncompetitive blocker, chlorpromazine, at its specific binding site (common to the five subunits) demonstrated that access to this site by chlorpromazine increases spectacularly in state A. These results agree with the hypothesis that this specific site is part of the ion channel. Chlorpromazine identifies the ion channel by affinity labeling.

What Is New with Channel Receptors?

The first amino acid of the active site was identified at the level of the α chain with an affinity label reacting to cysteins by Arthur Karlin and his colleagues in 1984. In fact, the situation is complicated. Our work with Jean-Luc Galzi in 1991 showed that the use of a more ubiquitous affinity label revealed that several amino acids enter into the composition of the two active sites present per receptor molecule at the level of three

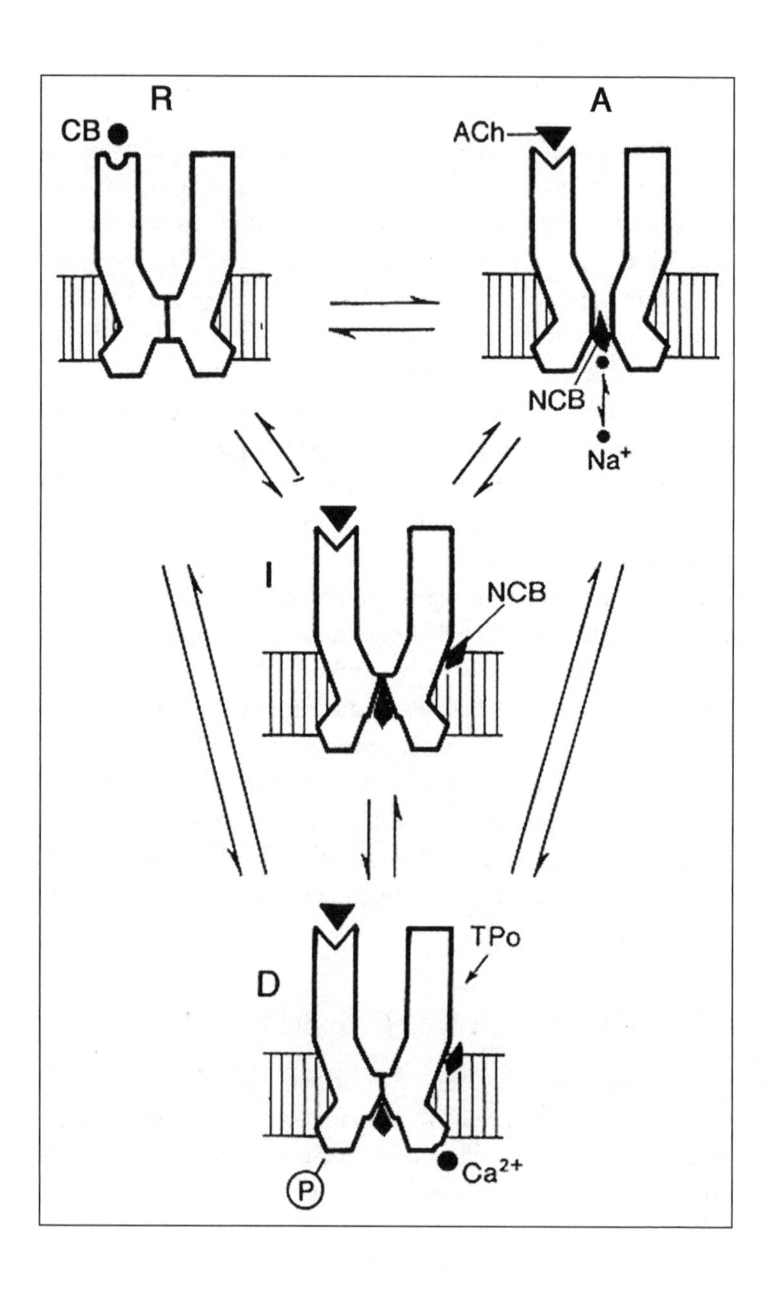
R
A
CB
ACh
NCB
Na+
I
NCB
D
TPo
P
Ca2+

loops of the α chain, A, B, and C. The principal labeled amino acids are aromatic and create an electronegative pocket, complementary to the quaternary charge of cholinergic ligands. A fourth loop, D, on the non-α chains participated in the structure of the two active sites, with two other loops, E and F (Figure 39). These data agree with the fact that in the case of the muscle receptor, these two sites are not pharmacologically identical and are situated, in the case of the muscle receptor, at the interface between two different subunits.

Covalent labeling of the high-affinity site of the channel blockers by chlorpromazine (Jérôme Giraudat et al., 1986) and subsequently by triphenyl methyl phosphonium (Ferdinand Hucho et al., 1986) identified the hydrophobic segment MII as forming the walls of the ion channel with a fivefold pseudosymmetry around the axis of rotation. Experiments with site-directed mutagenesis first performed by the teams of Shosaku Numa and Bert Sakmann, and then by ourselves in collaboration with Daniel Bertrand and Sonia Bertrand, confirmed the role of MII in ion transport. In particular, a ring close to the cytoplasmic side intervenes in ionic specificity and in the difference between inhibitory receptors, such as the GABA recep-

Figure 38: *Different conformational states of the acetylcholine receptor.* Prolonged exposure of the receptor to acetylcholine causes a reduction in amplitude of the response to the transmitter. Proposed models of desensitization involve several conformational states that bind various ligands with different affinities. R is the resting state, A is active with open channel, I and D are desensitized with closed channel. Affinity for acetylcholine is higher in I and D than in A. From J.-P. Changeux, "Functional Architecture and Dynamics of the Nicotinic Acetylcholine Receptor: An Allosteric Ligand-Gated Ion Channel," *Fidia Research Foundation Neuroscience Award Lectures* 4 (1990): 21–168.

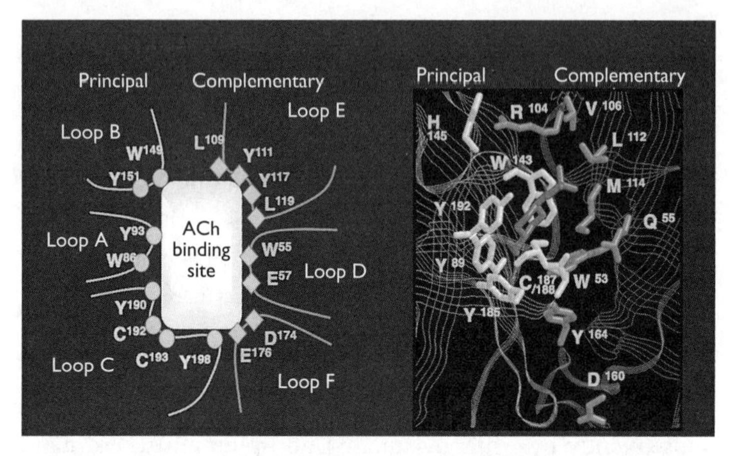

Figure 39: *The acetylcholine receptor site.* Biochemical and genetic data from *Torpedo* (J.-P. Corringer, N. Le Novère, and J.-P. Changeux, "Nicotinic Receptors at the Amino Acid Level," *Annual Review Pharmacology and Toxicology,* 40 [2000]: 431–458) and structure from *Lymnaea* (Brecj et al., "Crystal Structure of an ACh-Binding Protein Reveals the Ligand Binding Domain of Nicotinic Receptors," *Nature,* 411 [2001]: 269–276). From J.-P. Changeux and S. Edelstein, *Nicotinic Acetylcholine Receptors* (Paris: Odile Jacob, 2005); reprinted with permission.

tor, whose channel is selective for chlorine anions, and excitatory receptors, such as the nicotinic receptor, the channel of which is selective for cations.

The distance between binding sites of cholinergic effectors and the high-affinity site for channel blockers is 2 to 3.5 nanometers, of the same order as the distance between hemes in hemoglobin. In this respect the interaction between these two sites is an allosteric interaction, according to our 1963 definition. The presence of positive cooperative effects between binding sites of acetylcholine, the discrete or all-or-nothing character of the opening of the ion channel, and the oligomeric quaternary structure of the receptor molecule fall in the context of our 1965 model. Nevertheless, the pseudosymmetrical organization of the molecule with a single axis of rotation perpendicular to the membrane and its cascadelike interconversion through several conformational states confer unconventional properties probably related to the transmembrane situation of the receptor molecule.

We encounter most of these properties in receptors for $GABA_A$, glycine, and serotonin ($5HT_3$), which compose, with the nicotinic receptor, a superfamily of receptors coupled to ion channels. These receptors are also the targets of powerful pharmacological agents, such as benzodiazepines for the GABA receptor, and strychnine for the glycine receptor. A hypothetical schema of evolution of this family emphasizes first of all the acquisition of symmetry: the genesis of symmetrical oligomers ensures the formation of a transmembrane channel, cooperativeness, flexibility, and thus electrochemical transduction. In a second stage we note a partial rupture of symmetry by combination and formation of hetero-oligomers, which produces considerable functional diversification. A hypothetical common ancestor could have existed 2.5 billion years ago, in prokaryotes, where homologous cationic receptors have been identified. The oldest anionic receptor seems to be that of glycine. In both cases the formation of hetero-oligomers seems to have been late. It is at the origin of the differentiation of the benzodiazepine sites in the case of the GABA receptor. Taken together, these data on receptor channels support the hypothesis that the acetylcholine receptor and its homologues are bona fide allosteric membrane proteins, but they possess original properties of their own.

Receptors with Seven Transmembrane Domains

Receptors with seven transmembrane domains compose a vast family of molecules. Our knowledge about them comes on the one hand from work on bacteriorhodopsin of *Halobacterium halobium* and on the other from research on rhodopsin itself (see Figure 3), as well as transmitters and hormones coupled to G proteins.

Bacteriorhodopsin is a proton pump activated by light in the primitive halophyte bacterium *H. halobium.* It consists of a single twenty-six-kilodalton protein that contains a molecule of retinal. In the dark the retinal is in the all-trans form and establishes a Schiff base with the protein. In the light it isomerizes in less than ten picoseconds in the form of 13-cis, then reverts to its original form via a cycle with several intermediaries and a release of protons. The three-dimensional structure of rhodopsin was described by Richard Henderson and his colleagues using electron microscopy: it revealed a coiled folding of the polypeptide chain and seven transmembrane domains forming a bundle with a central pocket, where the retinal binds, and an axial channel for the protons. There is a conformational change in the photoactivation cycle that affects principally the most cytoplasmic part of the molecule. This consists of an allosteric transition in the true meaning of the term, since the distance between the retinal and the cytoplasmic sides is on the order of 1.2 nanometers.

Rhodopsin constitutes 90 percent of the proteins of the rods of the vertebrate retina (see Figure 3). It consists of a single chain of 348 amino acids, half of which are integrated in the membrane phase, coiled like bacteriorhodopsin with seven transmembrane domains. Photoactivation of rhodopsin causes allosteric activation of the heterotrimeric G protein (αßγ) (transducin), which interacts with the cytoplasmic side of the rhodopsin. It also leads to activation of guanosine monophosphate (GMP) phosphodiesterase and then a fall in intracellular cyclic GMP, which causes closure of the sodium channel, and the plasma membrane hyperpolarizes. Neurotransmitter and hormone G protein–bound receptors compose a large family of homologous proteins with seven transmembrane domains, including another type of acetylcholine receptor called muscarinic. One of the most important of these receptors is the ß2 adrenergic receptor, the DNA of which was cloned by Robert Lefkowitz and his colleagues in 1986, some four years after the acetylcholine receptor was cloned. In general, all these receptors present a profile similar to that of rhodopsin: seven hydrophobic membrane segments of twenty to twenty-five amino acids. In the case of the ß2 adrenergic receptor, most of these domains participate in the formation of the pocket in which the transmitter fits, mainly concerning domains II and VII. In the case of peptide receptors, the binding sites involve loops and extracellular amino acids. Signal transduction is accomplished by a conformational cytoplasmic transition that ensures allosteric coupling between the transmitter binding site and the site of interaction of the G protein.

The 1965 concerted model of allosteric transition by Monod, Wyman, and Changeux can apply to channel receptors involving a cascade of conformational transition R–A–I–D between active and desensitized states. The case for receptors with seven transmembrane domains is less obvious unless we consider the single receptor monomer as formed by seven α-helical subunits. But it has also been shown to associate with itself in the membrane as a labile oligomer. Interpretation of the functional properties of G protein–coupled receptors in the context of the allosteric model supposes the existence of several conformational states R, R*, and R*G. There would also be spontaneous interconversion, in the absence of ligand, of R to R*, which the pharmacological ligands would selectively stabilize. The mass of results obtained for channel receptors or G protein–coupled receptors is consistent with the hypothesis that they are allosteric membrane proteins: they illustrate their diversity and functional richness. Additional work is necessary.

Cellular and Molecular Mechanisms of Learning

Learning can be defined as any stable modification in behavior or psychological activity attributable to experience. It has a wider meaning than memory, which relates to availability and use of learning. Indeed, memory is the property of storing information but equally that of recognizing and recalling it. Learning and memory of course both involve the formation of a stable track, or engram. We speak of the memory of a computer as a component capable of storing and recalling data. In psychology, memory is related to the state of consciousness or attention. In 1890 William James called memory knowledge of an old psychic state reappearing in consciousness after having disappeared. Such secondary memory could be distinguished from immediate or primary memory, in the sense of being limited to a few seconds. More recently, these primary and secondary memories have been qualified as short- and long-term.

The philosophical problems of learning and memory involve two main themes regarding the origin of ideas or, more generally, of knowledge, in the general context of relationships with the outside world: the so-called mind-body problem. Monist hypotheses consider that all is reducible to unity, either to the mind, to a spiritualist monism, or to matter, a materialist monism. Dualism postulates two essentially irreducible principals, mind and matter. Spiritualist monism, defended by Bishop Berkeley (1685–1753), has little place in modern times, but dualist concepts, from Plato to John Eccles (1903–1997) and Descartes to Bergson, are still advocated in various forms, including total independence of body and mind, synchrony, parallelism, or interaction between them. In spite of its aspect of compromise, the interactionist concept has been considered throughout Western philosophical history. Of course, the essential problem remains the meeting point where this hypothetical interaction between body and mind might occur, maybe the pineal gland of Descartes, the striatum of Thomas Willis (1621–1675), or the right hemisphere of Eccles. In *Matter and Memory* Bergson wrote, "It follows that memory must be, in principle, a power absolutely independent of matter. If, then, spirit is a reality, it is here, in the phenomenon of memory, that we may come into touch with it experimentally. And hence any attempt to derive pure memory from an operation of the brain should reveal on analysis a radical illusion." These concepts are difficult to defend today, and they are naturally opposed by materialist monism, which took on a precise form in nineteenth-century Germany with Ernst Brücke, Hermann Helmholtz, and Emil du Bois-Reymond and their famous physicalist oath, according to which only physical and chemical forces act in the organism, the only authentically scientific task being to discover the specific mode of form of the action of these forces.

Evolution of Theories and Models of Learning: Birth of Experimental Psychology

In recent years the study of learning has been marked, apart from the general progress of neuroscience and informatics, by theoretical contributions from immunology, molecular biology, and physics. In a 1967 landmark article, "Antibodies and Learning: Selection versus Instruction," Niels Jerne suggested extending to the nervous system the selective models that had proved correct for the synthesis of antibodies. Especially in the human nervous system, each individual displays plasticity in his aptitude to learn, much like phylogenetically developed instincts. Learning is based on diversity of a fraction of DNA or its translation into protein, which controls the effective synaptic network, itself the substrate of instincts. Jerne wrote, "I would, therefore, find it surprising if DNA were not involved in learning, and envisage that the production by a neuronal cell of certain proteins, which I might call 'synaptobodies,' would permit that cell to enhance or depress certain of its synapses, or to develop others." Nevertheless, he forgot that selection of an antibody affects the differential proliferation of lymphocytes of the right genetic recombination, whereas in man neurons have practically ceased to divide before learning occurs.

So I proposed that selection might act not on genetic variations but on combinations of synapses and might therefore intervene as a regulatory factor of gene expression, or epigenesis. A genetic envelope would determine the rules of proliferation, differentiation, growth, and stabilization of neurons and synapses during development. At a sensitive stage of development connectivity would reach a maximum and consequently its full diversity. The activity of the network, spontaneous or evoked, would selectively stabilize certain synaptic combinations, whereas the others would regress. The mathematical formulation of this Darwinian model led to the formulation of a variability theorem, according to which the same input-output relationship could be obtained after learning by stabilization of distinct connectional networks. In 1987 Edelman extended the Darwinian model not only to development or even to learning, but also to higher brain functions. According to him, a functional unit is a group of 50 to 10,000 active neurons connected in a great variety of ways. A primary repertoire is prespecified during development. So selection brings into play a function of recognition with a threshold, and it retains from among the combinations of the primary repertoire those having a higher probability of later recruitment for a given distribution of input signals. Soon after, I proposed a model of learning by selection that concerns cooperative assemblies of active neurons (mental objects). The formal description of these states of activity depends on methods of statistical mechanics (William Little, 1975; John Hopfield, 1982; Pierre Peretto, 1983) and included Hebb's rule that synaptic efficiency changes when pre- and postsynaptic activities coincide in time. This model postulated that interaction with the outside world through a percept leads to selection by resonance from a great number of prerepresentations, variable in time and space; hence its qualification as neuronal Darwinism. I later took up this model with Stanilas Dehaene.

Allosteric Receptors and Molecular Models of Learning

The fundamental mechanisms of learning can be sought at several levels of organization. First, at the molar level is that of cooperative neuronal assemblies, and even Horace Barlow's cardinal neurons in charge of large populations of neurons. Also at the cellular level is that of the number of neurons and their connections, of the efficacy of the synapses they receive as well as their capacity to fire electrical impulses. And finally at the strictly molecular level is that of regulatory proteins capable of integrating several communication signals in time and space.

The acetylcholine receptor offers a particular example of an allosteric protein suitable for the elaboration of elementary models of learning at the postsynaptic level. As we have already discussed, its pentameric structure contains the ion channel and all the elements necessary for regulating its opening as well as the primary binding sites for acetylcholine carried, at least partially, by α chains. The molecule also binds allosteric effectors such as noncompetitive blockers and calcium ions at sites different from the precedent. High-resolution patch clamp recordings, and the chemical kinetics of fast binding of agonists or noncompetitive blockers, demonstrate the interconversion of the receptor molecule between several discrete conformational states, some of which are accessible spontaneously in the absence of an agonist. As we saw earlier, the main states are rest (R), active (A), intermediate (I), and desensitized (D). Steric and allosteric effectors of the receptor affect these transitions. Electrical fields affect the kinetics of desensitization (see Figure 38). The existence of these effects suggests that a second-order regulation could intervene in the opening of the ion channel by acetylcholine. The presence of several distinct sites on the same molecule enables topological conversion. The slow time scale of transitions introduces time constraints and ensures integration. Several signals, such as electrical fields and calcium, to which the receptor molecule is sensitive, can therefore serve as indicators of activity of neighboring synapses and, of course, of the postsynaptic neuron itself.

Our model of learning at the synaptic level makes use of the existence of transmembrane allosteric proteins, such as transmitter receptors and ion channels, capable of existing in at least two distinct states, such as an ion channel or enzymatic activity, and with a fixed subcellular localization. The presence of transmitter and of various physiological signals in the neighborhood of the postsynaptic site regulates the conformational equilibrium between the two states and therefore the efficacy of the biological response produced locally by the effector (see Figure 15). The formal model has served for computer simulation of homosynaptic regulation by differential stabilization of state R (facilitation) or D (depression) by the released effector. Models of heterosynaptic regulation, depending on cross-stabilization of a receptor by physiological (e.g., allosteric modulatory) signals in other synapses, have also been considered. The schema of classic conditioning is a particular case of heterosynaptic regulation where the temporal coincidence of two signals converging on the same allosteric state creates a synergic effect on the transition to this state (see Figure 15). This molecular model can serve to establish associative links between neurons, empirically as well as selectively. In the latter case the spontaneous activity of the postsynaptic neuron, an essential component of prerepresentations, intervenes in the regulation of allosteric equilibrium. Conditions of resonance can be defined.

Aplysia, a Cellular Model of Learning

Aplysia, the sea slug, lends itself to research on learning for several reasons. Its nervous system is simple: five pairs of ganglia and some 20,000 neurons. It contains identifiable neurons and groups of neurons, about 55 of which per abdominal ganglion have giant cell bodies about a millimeter in diameter and are therefore easily recognizable. They are localized similarly from one individual to another and have the same pattern of spontaneous discharge, range of sensitivity to transmitters, synthesis of transmitter, and connectivity. There are elementary behavioral patterns, the neural bases of which are found in identifiable neurons or groups of neurons, as described by Angélique Arvanitaki in 1942 and by Ladislav Tauc, Hersch Gerschenfeld, and Eric Kandel in the 1960s. Kandel distinguished four categories of behavior: reflex acts such as retraction of the siphon and gill, which is a graduated response subject to learning; fixed all-or-nothing acts, resistant to learning, such as ink release; complex motor behavior such as locomotion and feeding; and higher behavior involving social interaction, such as mating.

Siphon or gill retraction is triggered by mechanical stimulation, such as a jet of water on the siphon wall. It involves groups of sensory neurons and motor neurons identified in an abdominal ganglion, L7 for example, the unit activity of which is directly related to triggering and maintenance of the reflex, which is regulated homosynaptically. Repeated siphon stimulation leads to reduced amplitude of the response, or habituation, which results from modification of the response neither of the sensory neurons nor of transmitters, but reduces efficacy of sensorimotor synapses. The effect is strictly presynaptic; it results from a reduced number of quanta of transmitter released without a change in the size of the basic response to a single quantum of transmitter. Recordings of action potentials in the sensory neuron, the axonal terminals of which were modified, revealed that they were shortened in the presence of the potassium channel blocker tetraethyl ammonium. According to Kandel and his colleagues, this shortening was due to reduced ingoing calcium currents, which causes reduced transmitter release at the terminal.

Habituation of the gill reflex is subject to spontaneous extinction, which can be accelerated considerably by stimulation of the animal's head. This dishabituation involves another type of synapse from interneurons L28 and L29, which are themselves contacted by sensory neurons from the head. These synapses probably contact "habitual" sensorimotor terminals and result in an increase in the number of quanta released by increasing calcium entry. The regulation of sensorimotor synapse efficacy is thus controlled by calcium entry. Further, the effect of dishabituation, or heterosynaptic facilitation, can be simulated by application of serotonin or injection of cyclic adenosine monophosphate (AMP) in the sensory neuron. The cyclic AMP can be replaced by the catalytic subunit of cyclic AMP-dependent protein kinase, whereas injection of a specific inhibitor of protein kinase in the sensory neuron has the opposite effect.

So, Kandel's model was:

serotonin → increased cyclic AMP →
protein kinase activation → increased calcium currents →
increased transmitter release

The substrate of cyclic AMP-dependent protein kinase is a particular population of potassium channels responsible for a so-called S current. Recordings of single channels in isolated membrane fragments by patch clamp by Steven Siegelbaum and his colleagues indeed showed that the number of these channels decrease either by serotonin-sensitive adenyl cyclase stimulation or, in vitro, by the addition of kinase.

Kandel and his colleagues extended this analysis to an experimental situation that recalled classic Pavlovian conditioning. The conditioned stimulus was gentle tactile stimulation of the siphon, provoking a moderate retraction. The unconditioned stimulus was strong stimulation of the tail, which caused vigorous retraction of siphon and gill. The combination of these two modes of stimulation caused an approximately threefold increase in amplitude of the conditioned response. This effect was due to a prolonged action potential in the sensory neuron, which then caused increased transmitter release and response amplitude. Kandel's molecular model for explaining the effect of the temporal coincidence of the conditioned and unconditioned stimuli was shown in Figure 16. The synapse of the hypothetical serotonin interneuron with the sensorimotor neuron was the target of the effect. The serotonin-sensitive adenyl cyclase had its activity increased by the temporal coincidence of serotonin release and the action potential in the axon of the sensory neuron. This increased activity resulted in calcium entry associated with the action potential. The receptor-cyclase complex alone integrated in both space and time the convergent signals from the interneuron (unconditioned) and the sensory neuron (conditioned). This model, at least formally, agrees with our model of allosteric regulation, although, curiously, Kandel did not examine this aspect of learning.

Following this still largely interpretative model, the real stage of learning, which would satisfy the exigencies of topological convergence and temporal coincidence, would be at the postsynaptic level of an axo-axonic synapse. Regulation of transmitter release that resulted would be secondary and would, in a way, constitute a step in the "read-out" of initial learning. Kandel's cellular model of learning has the merit of involving simple molecular mechanisms (for instance, cyclic AMP and protein kinases), but these mechanisms may not be as generally applicable as hoped for. There exist many other second-messenger systems that could intervene in cellular learning of this nature. Furthermore, we still do not know about all the neurons and circuits involved in this relatively simple learning. Other forms of regulation could of course occur in higher animals, thus bringing into play systems such as the cerebellum and hippocampus. The model proposed here is strictly empiricist in nature. It is only indirectly related to Pavlovian conditioned reflexes, in that the conditioning leads not to a qualitative change but only to a quantitative change in the response. The human brain cannot be reduced to a simple collection of thousands of *Aplysia* ganglia.

The Chemistry of Consciousness

THE CHEMISTRY OF WAKING AND SLEEPING

Our brain contains nerve cells, themselves made of molecules, and synaptic communications between nerve cells involve chemical signals. We know that our state of consciousness changes with waking and sleeping and that we can precipitate the transition toward sleep with sleep-inducing chemical agents. The surgeon can put us to sleep by injecting a general anesthetic: we become unconscious. So how does chemistry influence those neural systems that govern our state of consciousness?

We owe to Henri Piéron, working at the beginning of the twentieth century, the first concepts of the chemistry of sleep, which he compared to the effect of drugs on wakefulness. In his remarkable experiments he managed to transfuse hypnotoxins in the blood of a dog deprived of sleep to a receiver dog, in which signs of deep sleep appeared. Thus, there must be chemical factors in sleep. In 1892 Goltz had shown that ablation of the dog's cerebral cortex did not alter sleep-wakefulness rhythms. In the same way, human anencephalic "monsters" alternate between sleep and waking, as well as between crying and smiling. So there must be noncortical centers for consciousness. In 1936 Frédéric Bremer sectioned the brain stem and then recorded the EEG. He found that if the section was anterior, there was permanent slow-wave sleep, but if it was posterior, there was a normal alternation of waking and sleep. So, if waking needs the brainstem to be intact, is that due to the need for sensory afferents? To answer this, Giuseppe Moruzzi and his colleagues sectioned the brainstem laterally, thus cutting the sensory pathways, but this had no effect. On the other hand, a median section abolished the waking-sleeping pattern, and the EEG reading remained sleeplike. They deduced that there is a reticular formation in the brainstem that intervenes in waking, independent of sensory afferents, and which projects anteriorly to wake the cerebral cortex. In 1958 Cesira Batini and Moruzzi performed other sections in the cat that caused either permanent sleep or waking. They concluded that there exist discrete regulatory centers in the reticular formation for each state. In the 1970s Michel Jouvet demonstrated the contribution of cholinergic and noradrenergic neurons during waking, and he suggested that there are transmitters concerned with states of sleep and wakefulness. In 2005 Barbara Jones proposed that rather than discrete waking or sleeping centers, the neuronal systems for these states were distributed through the brainstem, using different transmitters, or neuromodulators. She considered glutamate, abundant in the ascending reticular formation and in nonspecific thalamocortical projections, the main waking transmitter, essential for cortical wakefulness and its accompanying muscle tone.

Noradrenaline is also an important promoter of waking, liberated by the locus coeruleus, which is active during that state, less so during slow-wave sleep, and inactive during paradoxical sleep. Its neuronal discharges are maximal during attention, orientation, and stress, which stimulate the sympathetic system. Agonists of the α1-adrenergic receptor provoke sleep by closing potassium channels of thalamocortical neurons, thereby causing depolarization and cortical activation. An opposite effect results from α2-adrenergic receptors. Dopamine, like noradrenaline, is a promoter of waking, but the activity of dopaminergic neurons of the substantia nigra and the ventral tegmental area do not vary with state except in terms of rhythm: bursts during waking or positive reward and tonic discharge during sleep. Drugs that increase the effect of dopamine, such as amphetamines and modafinil, stimulate cortical activity and waking much as noradrenaline does. Histamine stimulates waking. Histaminergic neurons in the tuberomamillary nucleus of the posterior hypothalamus discharge during waking, whereas antihistamines cause somnolence.

Serotonin causes calm wakefulness with a feeling of satiety, just before the onset of sleep. It is released by neurons of the raphe at the level of their ascending projection to the forebrain or descending to the spinal cord. Lesions of these neurons cause insomnia, but stimulation does not cause waking. Their spontaneous activity increases during waking and decreases during slow-wave sleep. Drugs that decrease the effect of serotonin cause insomnia and an increase in eating, sexual behavior, and aggression. Acetylcholine stimulates cortical activity during waking and paradoxical sleep: lesions of cholinergic neurons in the forebrain decrease cortical activity and attention. Drugs that cause inhibition of muscarinic receptors induce slow-wave sleep and hinder attention and memory. Inhibitors of the nicotinic receptor can act as general anesthetics, whereas agonists, like nicotine, increase wakefulness and vigilance. Certain peptides of the middle and posterior hypothalamus, such as orexin and hypocretin, stimulate waking and eating, and their absence, due, for instance, to a genetic deficit, causes narcolepsy. Their function may be to stimulate the locus coeruleus and histaminergic and cholinergic neurons.

GABA is the most widespread inhibitory transmitter of the CNS. It is present in inhibitory neurons, both interneurons and those with long projections. It imposes a slow rhythm by inhibiting excitatory systems, and it contributes to the establishment of oscillatory activity in the limbic system and neocortex. GABAergic neurons block sensory input to the thalamus, particularly during slow-wave sleep. They also buffer cortical activity by long connections from the basal forebrain. As they locally inhibit cholinergic neurons, they control states of waking and sleep. They also help control the sympathetic system. Drugs that facilitate GABAergic transmission, such as pentobarbital, are general anesthetics, whereas benzodiazepines diminish cortical activity and encourage slow-wave sleep. Other sleep mediators, such as somatostatin, corticostatin, and adenosine, facilitate slow-wave sleep.

Acetylcholine is the main transmitter of paradoxical sleep. It is present in neurons of the pontomesencephalic tegmentum, which discharges during waking and, most of all, paradoxical sleep, and the destruction of which leads to loss of paradoxical sleep. Carbachol injected in the oropontine reticular formation acts on nicotinic and muscarinic receptors to cause paradoxical sleep. Noradrenaline, serotonin, and histamine play a permissive role in paradoxical sleep: neurons containing these mediators cease their activity before and during paradoxical sleep. They act reciprocally with cholinergic neurons during sleep-wakefulness cycles. Dopamine is released during paradoxical sleep but not during slow-wave sleep. Could that correspond to emotional activity and hallucinations that accompany dreams? GABA blocks sensory input in paradoxical sleep, when its release is maximal in the locus coeruleus and raphe. GABA also contributes to the muscle atonia of paradoxical sleep.

All these examples show unambiguously that there exists a complex neurochemistry of physiological regulation of states of waking and sleeping.

GENERAL ANESTHETICS

Before the nineteenth century surgery was performed on the patient while he was awake. Alcohol and morphine were used to diminish pain, but they were inefficient and dangerous in surgery. The first general anesthetic synthesized was nitrous oxide—laughing gas—by Joseph Priestley in 1769. He was an English clergyman, later accused of sorcery. In 1799 Humphrey Davy tried nitrous oxide on himself and several other people, and he noted that it caused analgesia and loss of consciousness. It was used in side shows in itinerant circuses. Davy suggested it could be used in surgery, but his idea was not followed up: the redemptive value of suffering was popular in Christianity at that time. Michael Faraday (1791–1867) discovered the narcotic effect of ether vapor and used it recreationally. But it was dentists who first used general anesthetics systematically, employing nitrous oxide or ether. The term *anaesthesia* comes from Oliver Wendell Holmes in 1846. Chloroform was introduced by James Simpson in 1847 as an aid in obstetrics, but the clergy denounced it as the work of the devil. Nevertheless, in 1853 Queen Victoria delivered her seventh child under the effects of chloroform (Figure 40). In addition to general anesthesia by inhalation of gas, such as nitrous oxide, or volatile substances like ether or chloroform, unconsciousness can be induced by intravenous injection of, for example,

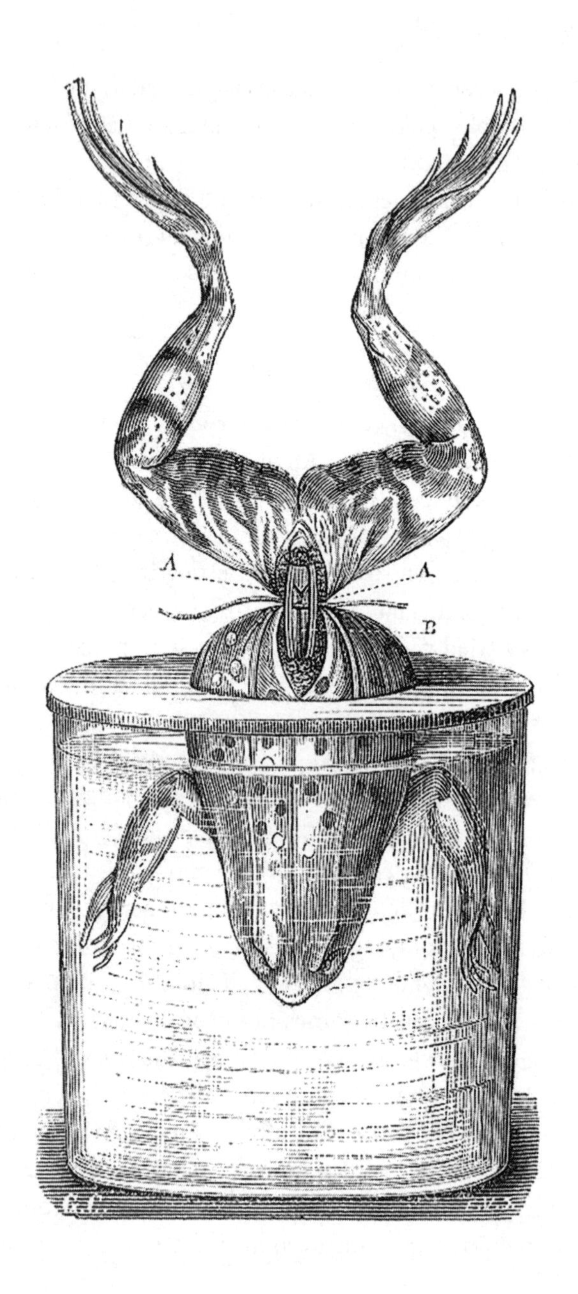
A
A
B

pentobarbital. General anesthesia must be distinguished from local anesthesia, in which there is local or regional loss of pain sensation but not of consciousness. We must also distinguish anesthesia from analgesia, which is a loss of perception of pain.

Ether was the first volatile anesthetic to be used systematically in surgery, but it causes secondary effects like postoperative nausea, vomiting, and respiratory irritation. And it is explosive. Halothane was much used later. It is not explosive and has fewer side effects, although it can cause respiratory or cardiac arrest and be toxic to the liver. Nitrous oxide is odorless and nonexplosive but less effective, so it is normally used with another anesthetic. Some injectable agents, notably barbiturates, are used for induction of anesthesia. Others are basic anesthetics, such as ketamine, which is related to phencyclidine and can cause delirium and hallucinations. Diazepam is a sedative anxiolytic acting on GABA receptors. Again, these substances are usually employed along with other anesthetics, analgesics such as opiates, neuroleptics such as chlorpromazine, antihistamines such as promethazine, or curare-like drugs to relax muscles.

Cerebral electrical activity (the EEG) can be recorded with electrodes on the scalp, and it varies greatly between sleep and waking, but also during general anesthesia. There are three levels of anesthesia. In level 1 patients respond to verbal instructions. In level 2 they are semiconscious, responding only slowly and with blurred speech. Level 3 brings full unconsciousness. As the patient progresses from level 1 to 3, the

Figure 40: *Claude Bernard and general anesthesia.* This drawing from Claude Bernard, *Leçons sur les anésthésiques et sur l'asphyxie* (1874), shows that general anesthesia by chloroform in the jar is mediated by nerves (A) and not blood vessels (B), which have been ligated in the body of the frog. Collection of J.-P. Changeux.

EEG shows increasing sigma waves at 10 to 15 hertz, and at level 3 gamma waves, at 30 to 60 hertz, are decreasing.

In 1999 Pierre-Olivier Fiset and colleagues observed that during general anesthesia (with propofol), cerebral blood flow measured by PET decreased globally by 20 percent. When the concentration of anesthetic was increased, there was decreased activity in the thalamus and various cortical areas. On the other hand, there was increased blood flow in certain other cortical areas and the cerebellum. These experiments also showed reduced blood flow in the ascending midbrain reticular formation, a structure known to be involved in regulation of level of consciousness. Work by Steven Laureys in 2005, and others, confirmed a significant down-regulation of the network involved in the conscious neuronal workspace.

MECHANISMS OF ACTION ON THE MEMBRANE

Four types of mechanism have been suggested.

A first mechanism involves the physical state of the membrane. In 1899 Charles Overton and Hans Meyer independently suggested that general anesthetics modify the physical state of the cell membrane, particularly its lipids. Overton determined the minimal concentration of anesthetic substances required to produce irreversible immobilization of tadpoles in a bowl and compared that concentration with the solubility of the test substances in lipids. Meyer concluded that narcosis begins when a chemically indifferent substance reaches a certain molar concentration in cell lipids. This relationship between anesthetic activity and lipid solubility was later confirmed.

A second mechanism involves the volume of the membrane phase. Keith Miller and colleagues in 1973 showed that newts immobilized by general anesthetic recovered their movements when a hydrostatic pressure of 100 atmospheres was applied, which is consistent with this theory. Quantitative analysis suggested that anesthesia happens when the volume of the lipid phase increases by about 0.4 percent. This change in volume provoked by anesthetics might modify conformational transitions of membrane proteins that contribute to electrical activity of neurons.

Another mechanism involves the fluidity of the membrane phase. In general, spin resonance spectroscopy demonstrates increased fluidity of membrane lipids by general anesthetics, but at much higher concentrations than those that are effective for anesthesia.

Finally, mechanisms involving the formation of hydrates have been suggested. In 1961 Linus Pauling and Keith Miller independently suggested that general anesthetics "freeze" water molecules in the form of an anesthetic-hydrate complex on the surface of the cell membrane, which interferes with its physiological properties. The correlation between the formation of hydrates and pharmacological effects was much less close if there was lipid solubility.

RECEPTORS FOR GENERAL ANESTHETICS

The theory of nonspecific effects of general anesthetics has been put in doubt by the observation of stereoselectivity in their mode of action. For example, the S(+) optical isomer of isoflurane is 50 percent more effective than the R(−). In the same way, some isomers of pentobarbital are twice as active as others. This phenomenon has been reported for many anesthetics. Several sites of action are possible (Nick Franks, 2008). General anesthetics are known to inhibit sodium, potassium, and calcium channels, but the necessary concentration is generally much higher than that effective for anesthesia. The only exception is the inhibitory synaptic potassium current, which is reversibly activated by weak concentrations of general anesthetic (Franks et al., 1988). This stereoselective effect on a homologous channel activated by anesthetics was also described recently by Michel Lazdunski in mammalian neurons. Thus, this is one possible target site.

RECEPTORS LINKED TO ION CHANNELS

NMDA (N-methyl D-aspartate) glutamate receptors could be a privileged target for certain anesthetics, such as ketamine, phencyclidine, MK801, and conotoxin G, all of which act as reversible noncompetitive channel blockers. Hans Flohr suggested that their anesthetic effect could be due to their inhibiting the NMDA receptor. Acetylcholine nicotinic receptors could equally be blocked by general anesthetics at the ion channel. For instance, ketamine might bind at segment M2 of the channel. Affinity labeling by a radioactive general anesthetic by Jonathan Cohen and colleagues demonstrated a site connecting M2 to M1 and M3 within the transmembrane domain, so there could be a specific allosteric site for general anesthetics acting on allosteric transitions of the receptor (Figure 41). Local anesthetics are well known to bind at a high-affinity site located at the ion channel in the M2 transmembrane segments, where they could stabilize the receptor in a desensitized state. The inhibitory GABA receptor could also be activated by general anesthetics; it is even the most plausible hypothesis. Indeed, pentobarbital increases the affinity of the response of the GABA receptor for general anesthesia approximately threefold. Halothane and propofol could have the same effect. Analysis of results from voltage clamping suggests that

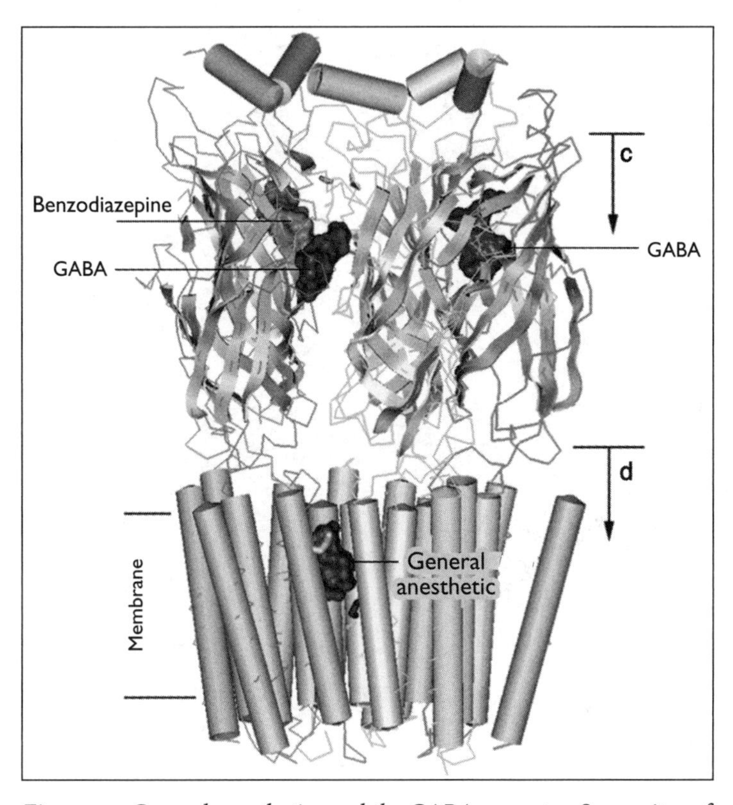

Figure 41: *General anesthetics and the $GABA_A$ receptor.* Some sites of pharmacological importance in a molecular model of the $GABA_A$ receptor. At the membrane level a site for a general anesthetic is shown. At the synaptic level the transmitter GABA is distinguishable from the homologous sites for benzodiazepine, an allosteric activator. From G. D. Li et al., "Identification of a $GABA_A$ Receptor Anesthetic Binding Site at Subunit Interfaces by Photolabeling with an Etomidate Analog," *Journal of Neuroscience* 26 (2006): 11599–11605; reprinted with permission.

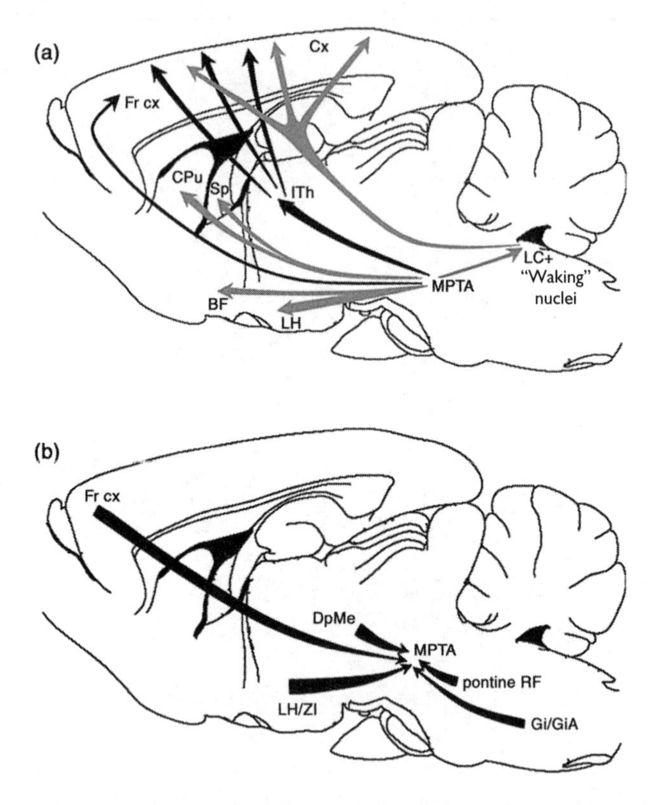

Figure 42: *A brain region for general anesthesia.* Devor and colleagues showed that local micro-injection of pentobarbital and other allosteric agonists of GABA receptors in the mesopontine tegmental anesthesia area (MPTA) of the rat causes reversible general anesthesia. The drawings represent efferent (a) and afferent (b) projections of this area. From I. Sukhotinsky et al., "Neural Pathways Associated with Loss of Consciousness Caused by Intracerebral Microinjection of $GABA_A$-Active Anesthetics," *European Journal of Neuroscience* 25 (2007): 1417–1436; reprinted with permission from John Wiley and Sons.

these anesthetics increase the frequency of opening of the chloride channel but do not act on basic conductance. Benzodiazepines have an effect different from that of general anesthetics, so there could be mutual potentiation with anesthetics. Studies of binding sites of general anesthetics with the $GABA_A$ receptor by Richard Olsen and Jonathan Cohen and their colleagues in recent years have identified a site common to these components (Figure 41): different anesthetics could bind at the amino acids in the M1, M2, and M3 membrane segments. They would prolong the effect of the transmitter and so modify the EEG, an essential part that would be interactions between inhibitory interneurons.

SPECIFIC NEURONAL CIRCUITS

Jingyi Ma and colleagues in 2002 suggested that reticulothalamic or septohippocampal systems could be the privileged target for certain general anesthetics. Marshall Devor's group in 2001 showed the existence in the rat of specialized neurons in the mesopontine region, specific pharmacological inactivation of which led to general anesthesia (Figure 42).

The overall results from studies of sleep-waking cycles and research on targets for general anesthetics thus show clearly that regulation of states of consciousness depends on complex chemistry involving multiple neurotransmitters within discrete neuronal arrays in which allosteric receptors play an important role.

VI

Where Do We Stand Today?

FROM *NEURONAL MAN* TO *THE PHYSIOLOGY OF TRUTH*

The Development of Brain Science: A Little History

Since my first lectures at the Collège de France more than thirty years ago, my aim has been clear: to take up the challenge of molecular biology and apply its paradigms and methods to a study of the brain and its most highly integrated functions, such as consciousness and thought. So where do we stand all these years later? Let us try to establish some new facts and some research perspectives for the decades to come.

Neuronal Man appeared in 1983, and its English version in 1985. It attempted to establish a pertinent causal relationship between structure and function, taking into account the successive levels of material organization in living organisms, from the elementary level of atoms and molecules to the highest levels, which in the brain means conscious thought.

In today's age of molecular biology all the cellular and

subcellular components of the nervous system, and the signals with which they communicate, can be defined in terms of molecules or organized molecular systems. It thus seems legitimate to extend to neurobiology the concept of "educated materialism," as Gaston Bachelard did for chemistry in 1953. In his famous *Testament,* Jean Meslier (1664–1729) wrote prophetically, "For the matter of a stone or heap of sand to begin to think, it would have to change, to be modified, to be transformed into an animal, a living man." We now envisage our brain as the synthesis of multiple "nested" evolutions: the evolution of man's ancestors at the level of the genome, the ontogenetic evolution of the embryo and its postnatal development at the level of neuronal networks, evolutionary dynamics of thought, and social and cultural evolution. This is so much the case that Thomas Huxley's lecture to the British Association in Belfast in 1874 seems amazingly up-to-date. He proposed that many of the best arguments concerning consciousness "applied equally to brutes and men." Consciousness is the result of molecular changes in the brain. We are like conscious automatons.

A Cartesian view of the nervous system as a connectional network and the desire to fabricate machines to rival the human brain began with Blaise Pascal's calculating machine in the seventeenth century. Around 1920 appeared the first electronic calculators, followed in the 1960s by supercomputers. Alan Turing and John von Neumann in the 1930s to 1950s were the main theoreticians. Turing asked, "Can machines think?" in his "Computing Machinery and Intelligence" in 1950, in which he described the mathematical theory of his famous Turing machine. In 1948 Norbert Wiener wrote *Cybernetics,* and he collaborated with the mathematician von Neumann, the physiologist Warren McCulloch, and the anthropologist Gregory Bateson. They developed the ideas of feedback, models, and systems. The model became a simple theoretical or mechanical representation of the brain and its functions. The system described interaction of constitutive elements forming a functional whole, the properties of which were more than the sum of the elements. Later, this cybernetic movement was joined by that of artificial intelligence, propounded by John McCarthy, Claude Shannon, and Herbert Simon. The objective was to write computer programs demonstrating mathematical theorems. In 1956, with *Logic Theorist* by Simon and Allen Newell, and in 1957, with their *General Problem Solver,* aimed at resolving such problems as translation of languages, chess playing, and decision making, the first attempts to model cognitive functions emerged.

In parallel with progress in brain science, mathematical and mechanical cybernetics, and artificial intelligence, a new, essentially behavioral discipline emerged in the 1950s, cognitive psychology. As a reaction to nineteenth-century German experimental psychological methods, which were based largely on introspection, an objective study of behavior developed that was based on "external" observation of an animal in its natural environment. This was notably due to John Watson and behaviorism, beginning in 1913. In 1948 Edward Tolman broke a taboo in introducing the notion of intention and cognitive maps in both human and nonhuman animals. In 1956 George Miller and Jerome Bruner published work on mental strategies of subjects confronted with cognitive tasks, including, for example, the famous definition of the "magic number seven," as well as the extent and limits of human cognitive function. Theoretical linguistics was given a boost in 1957 with Chomsky's *Syntactic Structures.* Generative grammar and general rules for language production led to the production of computer programs dealing with universal grammatical rules, which enabled a computer to speak, translate, and even "think." "A computer has hardware and we write software for it; the brain is the hardware and the mind the software," said Chomsky. A computer program sufficed to account for mental processes, according to Jerry Fodor in *The Language of Thought* in 1975, regardless of the type of machine, one with microprocessors or one with neurons. Many philosophers and psychologists followed suit. This decerebralization of brain function, which feels like a sort of ontological dualism, in fact marks the limits of the cognitivist approach.

A major event took place in 1971, the first meeting of the American Society for Neuroscience. Some 1,100 scientists met on that occasion. Today there are some 40,000 members, representing traditional neuroscience disciplines such as neuroanatomy, neurophysiology, neurochemistry, and neuropharmacology, but horizons have widened. This is typified by the massive representation of molecular biology, cognitive psychology, and brain imaging. Imaging has created a new bridge between psychology and the brain in the wake of major technical advances in computerized tomography (CT) by Godfrey Hounsfield and Allan Cormack, PET, and MRI scanning by Peter Mansfield and Paul Lauterbur.

Molecular biology was marked by the systematic analysis of the first behavioral mutants of *Drosophila* by Seymour Benzer and the mouse by Richard Sidman in 1967. In 1970 came the isolation of the first neurotransmitter receptor by our team at the Institut Pasteur. In 1973 Gunther Stent provided a molecular basis for Hebb's theories on learning. From then on it was legitimate to attribute a molecular basis to higher brain functions. The evolution of cognitive psychology, with its roots in neurophysiology, was marked by the emergence of neuropsychology, the study of the effects of brain lesions in the tradition of Broca by Luria, Geschwind, and Hécaen, but also of psychophysics, child psychology, and the study of animal behavior.

Neuronal Man appeared in that context, after neuroscience had established itself solidly. It was a sort of manifesto and offered a synthesis of the new field, dealing with molecular biology as well as mental objects, problems of consciousness, and the "substance of the spirit." Since then there has been major progress, such as the sequencing of various genomes, including the human one, the development of neurocomputational models of cognitive function, and, at last, a scientific study of consciousness.

The Power of the Genes

In *Neuronal Man* I spoke of the invariance of species specificity in the organization of the brain, but also of the variations imposed by genetic mutations in terms of, for example, cerebellar anatomy, and behavior in *Drosophila* and the cricket. I also pointed out the paradox of evolutionary nonlinearity between increased complexity in cerebral organization (and in the accompanying behavioral changes) and the apparent invariance of DNA content from mouse to man. To account for this paradox, I suggested a spatiotemporal model of gene expression based on those proposed by Monod and Jacob in the conclusions to the Cold Spring Harbor Symposium of 1961. Fifty years later their ideas have remained valid but have become enriched by many hidden treasures revealed by the complete sequencing of several eukaryotic genomes.

First we might consider the origin of animal life. On the basis of still incomplete sequencing of two cnidaria, the coral

Acropora and the sea anemone *Nematostella,* Ulrich Technau and colleagues proposed in 2005 that animal life has a single origin. Sponges appear to be the common ancestors of the cnidaria, which would be the common ancestors of deuterostomes and protostomes. The origin of sponges could be a hypothetical unicellular common ancestor, perhaps a choanoflagellate, colonial protozoa that still exist today. Despite this common ancestral nature of the genome of cnidaria, it still demonstrates genetic complexity and diversity superior to those of higher eukaryotes and of man. One finds genuine bacterial and plant sequences that were later lost in higher eukaryotes, as well as genes coding for regulatory proteins or transcription factors that we still find in higher eukaryotes. One can conclude that *loss* of genes has an important place in the evolution of genomes, though horizontal gene transfer may also occur.

Although the human genome has more than 3 billion base pairs, we know that there are only some 20,000 to 30,000 introns and exons. Coding exons represent only some 1.2 percent of our genome. Therefore, the majority of sequences are noncoding, but they are very diverse in nature. There are 20,000 inactive pseudogenes, a similar number to coding genes. Introns make up some 30 percent of the total sequences. Furthermore, the number of transposable elements in the human genome is colossal, some 44 percent of the total sequences. This includes 8 percent of genuine endogenous retroviruses, 33 percent of various transposable elements, and 3 percent of true transposons transposable by "cutting and pasting." These transposable elements seem to have played a major role in vertebrate genome evolution by being integrated in successive waves and causing mutations or chromosomal reorganization.

The coding sequences of the human genome have well-defined functions. Genes for housekeeping proteins, which constitute a large fraction of the genome in unicellular (46 percent in yeast) and multicellular eukaryotes (43 percent in a worm), intervene in the basic metabolism of the cell and its duplication. Multicellularity proteins, which help differentiate a worm from yeast, consist of intercellular signal transduction systems (such as epidermal growth factor, EGF), intercellular adhesion proteins (such as fibronectin), and transcription factors (such as hormone receptors and homeotic proteins). Gene families peculiar to vertebrates, including humans, include those concerned with immunity defense and the development of the nervous system. The "proteome of the human mind" contains proteins specific for the production, propagation, and transmission of nerve signals (opiate receptors are new, compared with the fly), cytoskeletal proteins (65 actin genes in humans, compared with 15 in the fly), and the development of connections (for example, ephrins and cadherins). Genes coding for transcription factors also take off explosively in humans.

The distribution of "mind genes" in the genome is still not clear, but we can say with certainty that, in general, structural genes are not distributed randomly in the human genome. Huib Caron and colleagues in 2001 reported clusters of highly expressed genes on chromosomes, separated by regions of low gene expression. Some of these clusters involved genes related to cancer, and others are related to the organization of the body (Hox) or blood proteins. Others could apparently influence the specification of certain regions of the brain, such as the hypothalamus and cerebral cortex (Wee-Ming Boon et al., 2004). A question arises: could there be coordinated expression of genes for brain function? Gill Bejerano and his colleagues in 2004 described ultraconserved elements in the human genome. They consist of more than 200 base pairs and are "absolutely conserved" from mouse to man. "These ultra-conserved elements of the human genome are most often located either overlapping exons in genes involved in RNA processing or in introns or nearby genes involved in the regulation of transcription and development." Could they be regulators, resistant to evolutionary change, determining essential structures for the organism, such as the brain? We just do not know, but we must follow this story.

Since publication of *Neuronal Man,* several major discoveries in molecular biology have changed and enriched our interpretation of the human genome and its expression. First of all there are genes coding for RNA, which is not translated into protein. We are familiar with antisense RNA, but interference RNA is different. It is double-strand RNA in which one strand associates with a protein, RNA-induced silencing complex (RISC), while the other pairs with target messenger RNA, which it cuts in two. This is another mechanism to "silence" gene expression.

Constitutive splicing and alternative splicing have been known since the discovery of introns in genes. Recent progress has come from detailed analysis of the enzymatic mechanisms involved and of the generalized nature of the mechanism. In humans there are more than 100,000 messenger RNAs, which is some four times more than the number of genes. This bears witness to the abundance of messengers resulting from alternative splicing. A remarkable case is that of neurexin, a synaptic protein that associates with neuroligin, and the gene of which could be the origin of 2,000 different mature messenger RNAs. More than 10 percent of hereditary diseases could be associated with mutations at the level of junctions between exons and introns. From worm to human the proportion of genes showing alternative splicing appears to have increased from 22 percent to 35 percent.

The Genetic Origins of the Human Brain

Work on genes and the brain often provokes hostile criticism; nevertheless, the universality of the human and his brain is to be sought in his genes. In 2001 Paul Thompson and his colleagues used MRI to compare ten pairs of monozygotic ("identical") twins and ten pairs of dizygotic twins, matched for age, sex, handedness, and social level. There was a perfect correlation in the distribution of gray matter (the layers of nerve cell bodies) of the frontal, sensorimotor, and language cortex in identical twins. The correlation was much less strong in nonidentical twins, although the language areas were about 90 percent correlated. So the power of the genes is important at the level of the macroscopic anatomy of the brain.

Today our knowledge of the genetic events at the origin of the human brain is still incomplete. Two recent observations are important, however. First is the identification of the genes for microcephaly, which offers precious information about the growth of the brain from our ancestors to ourselves. Microcephaly is a hereditary condition presenting as a reduction of the volume of the brain to a third of its normal value (around 400 cubic centimeters, the size of that of *Australopithecus*). It is found in 2 percent of neonates and includes two syndromes. The first is called high-functioning, in which the brain is small and has a reduced gyral pattern. The children walk late, at two years, and their language capacity is severely reduced. The genes ASPM (abnormal spindle-like, microcephaly-associated) and MCPH1 (microcephalin)—both homozygotic—

and SHH (sonic hedgehog; heterozygotic) cause the deficit. The other is low-functioning, and it is manifest by gross anatomical deficits, profound mental retardation, spasticity, and early mortality. Genes involved are ARFGEF2 or ATR. The biochemistry of the products of these genes reveals a surprise. ASPM and MCPH1 proteins regulate the proliferation of neuroblasts, the precursors of the neurons themselves. They are proteins associated with microtubules, and they control the number and orientation of cell divisions in the neuro-epithelium, and thus the size of the brain (Susan Gilbert et al., 2005; Jacquelyn Bond and Christopher Woods, 2005). A study by Steve Dorus and his colleagues in 2004 of the evolution of other genes expressed in the brain has shown the following. Central nervous system genes evolve more rapidly than housekeeping genes, from mice to primates. Developmental genes, such as those associated with microcephaly, are among the most rapidly evolving genes in the nervous system. Genes related to the "mind" evolved more rapidly in human ancestors than in any other line. Does this evolution continue today? Dorus and his colleagues think so. They suggested an acceleration in evolution for microcephaly genes from 37,000 years ago, corresponding to an explosion of symbolic behavior in Europe, and of ASPM 5,800 years ago, just before the foundation of the first town in the Near East (Michael Balter, 2005). Evolution is still going on . . .

Since publication of *Neuronal Man,* new possible human ancestors have been identified, among which is *Sahelanthropus tchadensis* by Michel Brunet and his colleagues in 2002. Its skull resembles that of a chimpanzee from the back, and that of *Australopithecus* from the front. It lived 6 or 7 million years ago in Chad and is therefore older than *Australopithecus* and *Ardipithecus* and almost contemporary with *Orrorin tugenensis.* An unresolved question is whether it was a direct ancestor of man, of the anthropoid apes, or of both. It is difficult to say. We are faced with an ever denser human ancestral tree: detailed genomics would be very useful.

Sequencing the chimpanzee's genome is not yet finished, but what we already know is very enlightening. There is only 1.2 percent genetic difference between chimpanzees and humans, that is, 18×10^6 base pair changes, due mainly, it seems, to relatively macroscopic chromosomal events. These include fusion of chimpanzee chromosomes 2p and 2q to make a single human 2p + q. But there are also translocations, inversions, and duplications. The hypothetical phylogenetic tree, based on sequence homologies, puts the divergence of man from chimpanzee 5 to 7 million years ago, and man from orangutan 12 to 15 million years ago, as the chimpanzee is closer to man.

One rather unexpected aspect of work on the human genome is the revealing of its great variability from one individual to another. The 0.01 percent of differences between individual sequences are of several types, according to Svante Pääbo in 2003. There are "en bloc" variations. Haplotype blocks of 5,000 to 20,000 base pairs make our genome a mosaic. Of 928 such blocks studied in Africa, Asia, and Europe, 51 percent were present on all three continents, 72 percent on two of them, and 28 percent on one only. In this last case, 90 percent were in Africa, where variability was greatest. Outside Africa one can encounter all the African variations, which points to Africa's being the cradle of humanity. *Homo sapiens* is supposed to have emerged 50,000 to 200,000 years ago from a population of about 10,000 individuals of African origin. Did Neanderthal man contribute to the genetic pool of modern man, as far as we can tell from mitochondrial DNA? Not according to Pääbo, but his view is contested. As to "ethnic" genetic variability, it corresponds to only 10 percent of global human genetic variability. Interindividual variability in terms of haplotype blocks is much greater, and the genome of a European can be much closer to that of an African or an Asian than to that of another European. Then there are structural variations. These smaller variations, short DNA sequences of less than 1,000 base pairs, encompassing insertions, deletions, inversions, and duplications, are another important source of variability. This is single-nucleotide polymorphism (SNP). One encounters about 10 million in human populations, that is, one nucleotide in 300 in an individual's genome. Apart from SNP, larger variations, 3 million base pairs or more, often visible microscopically after staining of the chromosomes, like trisomy 21, are also common. Human SNP or other variability, such as gene copy numbers, is responsible for individual differences in predisposition to disease and response to medication, as well as in learning.

The Proteome and Cerebral Morphogenesis: From One Dimension to Three

Every day human genomics reveals new candidate genes that might enable us to better understand the "genome of the mind." Let us now proceed to look at the organization of the body, and the brain in particular, which results from regulation of gene expression in time and space. Detlev Arendt and Katharina Nübler-Jung in 1997 showed how the transition from polychaetes with a ventral nerve cord to chordates with a dorsal nervous system could be reduced to extremely simple movements at gastrulation (the formation of the embryonic germ layers), which could be controlled by a small number of homeotic genes. They further emphasized the great similarity between the expression of developmental genes for the nervous system in *Drosophila* and mice, particularly for the longitudinal distribution of the gene product, segmental distribution being more variable. This conservation of the molecular organizational plan for the developing nervous system from fly to mouse also applies to humans. These concepts recall and expand those in *Neuronal Man,* but they still stumble over the mechanisms relating to the increasing complexity of the brain within this organizational plan. A special feature, but one of general importance, is the differential expansion of certain areas of the brain. The most obvious example is that of the cerebral neocortex in mammals compared with a phylogenetically older cortex. More discrete, but nonetheless important, is the expansion of the frontal cortex, from 3.5 percent in the cat, 8.5 percent in the squirrel monkey, 11.5 percent in the macaque, 17 percent in the chimpanzee, to 29 percent in the human, according to Joaquin Fuster in 1989.

A simple experimental model helps us understand the mechanisms involved, such as the role of homeodomain transcription factor EMX2 and paired box gene 6 (Pax6) in the specification of cortical areas in the mouse. These two proteins are expressed in the dorsal telencephalic epithelium with opposite rostrocaudal (anteroposterior) gradients. EMX2 is maximum caudally, and Pax6 rostrally. Mutant mice have been bred with each of these genes knocked out (deleted). The consequences of deletion of EMX2 on the differentiation of cortical areas were measured either with cadherin 8 for rostral motor areas or cadherin 6 for lateral somatosensory areas. As predicted, there was a posterior expansion of the rostral cortex. In the Pax6 knockout, on the other hand, the opposite happened: caudolateral areas invaded anteriorly. We might then conceive that genes of this type, Pax6 in particular, could regulate the relative extent of the frontal cortex (Kathie Bishop et al., 2000).

We know from the work of Korbinian Brodmann that cortical areas are defined by relative cytoarchitectonic homogeneity, which means that the distribution and features of their neurons at the microscopic level are similar for a given functional area, and that each area is delimited by a more or less visible border. So the question of expansion of a given area can be asked in terms of shifting a border. Our computational model of 1994 provided an answer with only a modest molecular cost. Its mechanism of expression was based on a diffusion gradient of an embryonic morphogen and the formation of genetic transcriptional borders, such as can be observed in the early stages of development in *Drosophila*. The position of the border would change as a function of the initial gradient of morphogen, but also of the affinity of the transcriptional complex that it formed with embryonic gene products and its efficacy. Quite modest changes in the structure of transcription factors could have dramatic consequences for the position of borders, and thus for increased complexity of the brain.

Epigenesis by Selective Stabilization of Synapses

The chapter of *Neuronal Man* that had the greatest influence was the one on epigenesis. I revisited it in *The Physiology of Truth* in 2002, and it is still valid. I should, however, comment on the use of the term *epigenesis,* which is used more and more by developmental molecular biologists in the sense of regulation of gene expression during embryogenesis, essentially at the level of the chromatin, the complex of DNA and protein that contributes to the organization of the chromosomes. In 1973 we (Changeux, Courrège, Danchin) introduced the term to express the mechanism of selective stabilization of synapses by much later activity during fetal and postnatal development.

This non-Darwinian selection contrasted with more innatist hypotheses of functional validation, such as that of Hubel and Wiesel, and the more empiricist one of activity-induced growth. The theory of variability, according to which different learning inputs may produce different connective organizations and neuronal functioning abilities—but the same behavioral ability—answers the criticisms of cognitivists like Fodor, for whom it would be pointless to try to link neurology and psychology because that relationship is so variable.

THE ROLE OF SPONTANEOUS ACTIVITY IN EPIGENESIS BY SELECTION

This very important aspect of our original theory of 1976 was emphasized in *Neuronal Man* in the section of the chapter on epigenesis entitled "The Dreams of the Embryo." Since then Lamberto Maffei and Carla Shatz have demonstrated high spontaneous activity in the fetal retina, propagated as a coherent wave that can be disrupted by a nicotinic agonist, epibatidine. With Maffei and Francesco Rossi, we showed in 2001 that mice in which the β2 subunit of the nicotinic receptor had been inactivated demonstrated the great importance of this spontaneous activity in the development of the visual pathways as far as the visual cortex. Such mutant mice lacked high-affinity nicotine binding and showed abnormal segregation of retinal projections to their target nuclei in the thalamus and midbrain (the LGN and superior colliculus), as well as physiological reorganization of the LGN in terms of retinotopy and receptive field properties. In the same group, Matthew Grubb and his colleagues in 2003 found simultaneous loss and gain of function, and Thomas Mrsic-Flogel and his colleagues in 2005 demonstrated abnormal mapping of visual space on the colliculus. So we concluded that "organized" spontaneous activity in the retina is necessary for normal development of visual pathways from retina to cortex.

EPIGENESIS OF DIFFERENT SENSORY MODALITIES

Another recent theoretical development stems from some remarkable experiments of cross-reinnervation between sensory pathways of different modalities, such as vision and audition. In the eighteenth century Claude Helvétius, Diderot, and the Encyclopedists were already asking such questions as how someone with a thousand fingers might view the world, or if the auditory nerve could be replaced by the optic nerve. Mriganka Sur and his colleagues in 1988 tried related experiments on newborn ferrets. They removed visual target nuclei, such as the superior colliculus and visual cortex, and saw that the retina then projected to the medial geniculate body (MGN) of the *auditory* thalamus. After a few weeks, visual axons synapsed in the MGN, which responded to *visual* stimuli. Even better, the auditory cortex, which received axons from the MGN, responded selectively to oriented light stimuli just like complex neurons in the visual cortex! The normal primary auditory cortex contains a one-dimensional representation of the cochlea, whereas the reinnervated auditory cortex contained a two-dimensional map of the retina (Anna Roe et al., 1990). Orientation specificity and long horizontal connections were similar in the reinnervated auditory cortex to those in a normal visual cortex. In 2000 Douglas Frost and his colleagues clarified discrimination of various visual stimuli in a learning test in the hamster, before and after reinnervation of the auditory cortex by visual neurons. The result was unambiguous: the reinnervated animals had behavior similar to that of normal animals. Clearly, removing a given sensory input causes a global reorganization of pathways in the brain in favor of the surviving sense.

SENSORY COMPENSATION IN THE BLIND

The blind have often been credited with exceptional ability in nonvisual sensory modalities. Diderot noted their superior tactile skill. Today functional brain imaging has revealed functional reorganization in the brains of blind people. In the 1990s André Goffinet and colleagues used PET scanning to study cortical activation in a blind subject who was reading Braille. There was activation of the primary and secondary visual cortex bilaterally, whereas in normal subjects these are inactive. Studies with evoked potentials and fMRI have confirmed these observations, which demonstrate a reinnervation of the cortex in the blind, in particular of the visual areas, by somatosensory tactile inputs. Further, Christian Büchel in 1998 and Leonardo Cohen in 1999 showed that there were differences between the congenitally blind and those who acquired the deficit. Cohen noted that if the blindness began after a critical age of between fourteen and sixteen years, the activation of the primary visual cortex was less than in subjects blind from birth. A vascular lesion in the visual cortex of a person blind from birth has been known to cause alexia for Braille, and covering the eyes of a normal adult subject for five days was reported to facilitate reception of tactile sensation in the visual cortex. So is there very fast growth of connections? Or is this an activation and amplification of already existing collateral connections? These questions are awaiting answers. Daniel Goldreich and Ingrid Kanics in 2003 studied quantitatively tactile acuity in the blind and demonstrated a level of performance expected of a subject of the same sex but twenty-three years younger. In 2003 Amir Amedi and his colleagues compared a cognitive task of word generation and a task of verbal memory involving recall of a list of nine abstract names after a week. They found that blind subjects performed pretty much like the sighted subjects in word generation, but somewhat better in verbal memory. What was remarkable was that the visual cortex was activated during the memory task. This was therefore not a case of global reorganization of the visual cortex but an "epigenetic" reorganization, specific to the blind, involving tactile reading and verbal memory.

GENETIC CONSTRAINTS OF CONNECTIONAL EPIGENESIS

The original theory of selective stabilization of synapses defined a genetic envelope, the collection of genetic determinants or constraints intervening in axonal guidance, including recognition of the target, synaptic adherence, and stabilization or elimination of initial contacts. One of the important areas of progress in the last few years has been the demonstration of pathological changes in connectional epigenesis brought about by disturbances of the genetic envelope. These could be at the origin of serious childhood disease. In particular, dyslexia, autism, and mental retardation associated with fragile X syndrome are produced by pathological changes in the genetic envelope that are related to synaptic epigenesis. Dyslexia appears as reading difficulties in childhood, and numerous candidate genes have been identified. A particularly interesting one, discovered by Katariina Hannula-Jouppi and her colleagues in 2005, is a homologue of an axonal guidance gene called ROBO, identified in *Drosophila* and which, in the mouse, results in disturbance to axons crossing the midline of the brain.

I spoke earlier about autism as a deficit in the development of social relations and affective contact in early infancy. It is known to be associated with a spurt in axonal growth in the brain between six and fourteen months of age. Thomas Bourgeron and his colleagues identified two loci on the X chromosome associated with autism that correspond to neuroligins 3 and 4. They are adhesion proteins present postsynaptically that associate with neurexins, proteins that we have already discussed and which can have very diverse forms through alternative splicing. I have proposed that changes in neuroligins that interfere with cortical synaptogenesis affect mainly long-axon neurons directly concerned in the conscious neuronal workspace. These neurons should be especially vulnerable because of the very high ratio of axonal volume to somatic volume. Fragile X syndrome is the commonest form of mental retardation, triggered by an anomaly at the end of the long arm of the X chromosome. The gene responsible was identified by Jean-Louis Mandel and his group in 1991 and is called *FMR1* (fragile X mental retardation). The coding region is 1,900 base pairs and the mutations responsible for fragile X result from the expansion of the triplets CGG hundreds of times, until function is lost. The final result is disturbance of the neuronal cytoskeleton, which is needed for synaptic stabilization. There are genetic constraints for synaptic epigenesis.

MOLECULAR BIOLOGY OF EPIGENESIS BY SELECTION

In *Neuronal Man* I discussed the work that was under way on the molecular biology of the neuromuscular junction. An important step was the identification of the acetylcholine receptor and the development of methods of studying it in embryonic life. The receptor is initially spread uniformly over the embryonic muscle fiber, but it progressively accumulates beneath the motor nerve terminal in the middle of the fiber to form a neuromuscular junction, and the extrasynaptic receptor disappears. There is compartmentalization of gene expression, essentially involving transcription. According to our 1987 model, the "first messengers" implicated were either neurotrophic factors (such as neuroregulin, CGRP [calcitonin gene-related peptide], or agrin) for the junctional domain, or electrical activity (and the entry of calcium, which accompanies depolarization) for the extrajunctional domain. These first messengers correspond to different chains of signal transduction, which in the end control distinct transcription factors, which themselves attach to different DNA elements present on gene promoters coding for subunits of the receptor. Experimental results in the last few years have largely confirmed this model.

Problems of Consciousness

This title is the same as that of a section of chapter 5 in *Neuronal Man,* entitled "Mental Objects." In it I tackled a question that was later popularized by Crick and Edelman and has led to much debate. It is not so much that of the neural correlates of consciousness as the neural *basis* of consciousness. I wrote: "At the level of integration we are now discussing, what we might call *consciousness* can be defined as a kind of a global regulatory system dealing with mental objects and computations using these objects. One way of looking at the biology of this regulatory system is to examine its different states and identify the mechanisms that guide the change from one state to another." I gave the example of hallucinations, then that of transition from waking to sleeping and the role of the reticular formation and its various systems of neuromodulators. I suggested that reciprocal pathways returning from the cortex to the brainstem contribute to integration between centers. "From the interplay of these linked regulatory systems, consciousness is born." In this way I was applying to consciousness the concept of reentry as used by Edelman for synaptic selection, an idea that he later took up for consciousness. I also mentioned the "calculation of emotion," with a citation from Jean-Paul Sartre, for whom "emotion is a form of existence of consciousness," which led me to emphasize the role of the emotions in the evolution of conscious calculations. I finally concluded, "Man no longer has a need for the 'Spirit'; it is enough for him to be Neuronal Man." This same point of view was taken up by Crick in *The Astonishing Hypothesis* in 1994.

The strategy that Stanislas Dehaene and I have since adopted for our scientific study of consciousness depends on the elaboration of models, as I first conceived them with Philippe Courrège and Antoine Danchin in 1973. These models aim to represent behavior or a mental process on the basis of a minimal, but realistic, neural architecture and distribution of activity, if possible, in mathematical terms. They also aim to establish causal relationships between a specific behavior or *subjective* mental processes and *objective* neural events that can be tested experimentally, from the molecular level to the cognitive level. All theoreticians, however, are conscious of the fact that, as a product of their own brain, even their best model can never give a complete and exhaustive account of reality.

This model-making strategy is not new. It follows, and extends, a long tradition at the Institut Pasteur, notably that of molecular biology. Since then it has been applied to ever higher levels of organization, beginning with models of epigenesis by synaptic selection, then the level of neuronal networks, and then their extension to large assemblies of neurons governed by rules of statistical physics. These last models possessed the interesting property, far beyond John Hopfield's classic model, of functioning as palimpsests. Instead of collapsing catastrophically when one added more and more memory, they developed a stable system in which only the most recent memories could be recalled, while older memories were slowly deleted. This global property of limited memory depended in fact on quite simple microstructural properties, such as the mean number of synapses per neuron, as we discussed with Jean-Pierre Nadal in 1986. A model followed to deal with learning temporal sequences, like birdsong. A simple mechanism of a synaptic triad imposed global order: it dictated a compulsory temporal sequence in the activation of two synapses. The model postulated a macroscopic architecture of three layers of assemblies of neurons linked by synaptic triads. Spontaneous production of prerepresentations of sequences passing from one group of neurons to another permitted selection by resonance with the input percept. In this way learning temporal sequences by selection is possible.

Modeling of cognitive function was also part of our project in 1989 and 1991 to construct a formal organism capable of delayed matching to sample tasks and Wisconsin card-sorting tasks, knowing that performance in these tasks depends on the integrity of the prefrontal cortex. All such tasks require a mental effort. (I might have said "conscious" effort, but out of prudence did not.) The basic structure of this formal organism relies on two principles. First is the distinction of two levels of organization: one basic sensorimotor level and a higher level, analogous to Shallice's supervisory attentional system, where a generator of diversity depends on groups of rule-coding neurons, the activity of which alternates from one group to another. Second is the intervention of reward (positive or negative reinforcement) neurons in the selection of a rule that fits a signal from the outside world that is, for example, imposed by the experimenter. In addition, our 1991 model permitted memorization of test rules, with the possibility of rejecting them through reasoning, the internal testing of current rules by self-evaluation, and the detection of coincidence between anticipated rules and reward signals by allosteric receptors at strategic sites. The Tower of London test also involves the prefrontal cortex and was part of our 1991 model, which proposed a synthesis of a hierarchical organization, an ascending evaluation system, and a descending planning system, permitting the organization of sequences of successive moves leading to the realization of a given aim.

The year after the formulation of this model, my thoughts turned toward the neural basis of consciousness and attention. I considered various definitions of consciousness. There was Lamarck's inner feeling of 1809, or Spencer's suggestion in 1855 of new grouped states intervening between more primitive ones, creating a consciousness independent of the outside world, a sort of mental *milieu intérieur* analogous to that coined by Claude Bernard. Then there was a high hierarchical level defined by Hughlings Jackson as "the least organised, the most complex, and the most voluntary," and Bergson's current of consciousness. More recently, Llinás, Crick, and Edelman, among others, have considered consciousness, but without proposing an explicit neurocomputational model. In *A Cognitive Theory of Consciousness* in 1988, Baars presented a psychological model in which he distinguished unconscious encapsulated processes and a conscious global workspace, but without a specific neuronal basis. In 1991 I suggested extending the Wisconsin task model to Baars's global workspace hypothesis. There followed a long period of reflection. Some years later we were able to propose a plausible model of a neuroanatomical basis for Baars's workspace. We defined a space for simulation and virtual action, where we evaluated aims, intentions, and programs of activity with reference to interaction with the outside world, but also inner intentions, self and personal history, moral norms, and social conventions, all internalized in the form of long-term memory traces. The fundamental hypothesis behind this model in 1998 was that pyramidal neurons of the cerebral cortex, which have long axons and are capable of linking different cortical areas, even in both hemispheres and often reciprocally, constitute the principal neural basis of the conscious workspace. Pyramidal neurons linking cortex to cortex are essentially, but not exclusively, found in layers II and III of the cortex, and they are notably abundant in the prefrontal, parietotemporal, and cingulate cortex, which could take part in a neuronal circuit within the conscious workspace. It is remarkable that cognitive tasks that involve consciousness mobilize this circuit, whereas it is not activated during unconscious processes, as shown by visual stimulation using masking and attentional blink paradigms. Further, this circuit is strongly inhibited in vegetative states, during general anesthesia, and in coma, as shown by Steven Laureys and his colleagues in 2005. The dynamics of access to consciousness have recently (between 2005 and 2007) been measured by EEG methods using the model of attentional blink by our colleagues Claire Sergent and Antoine Del Cul; they have shown a maximum amplitude at between 300 and 400 milliseconds. This relatively slow access took place suddenly and in an all-or-nothing fashion. Our 2003 model accounted for this threshold effect on the basis of return connections, from top to bottom, from neurons of the conscious workspace to primary and secondary sensory neurons. Comparison of theoretical and experimental data was encouraging, and I think this was the first successful attempt to model the connections leading to access to consciousness.

Any realistic model that takes account of the enormous number of cerebral neurons and their connections soon encounters the explosive nature of the number of possible combinations. The brain contains systems that code certain combinations; they are acquired by learning. Hence the epigenetic rule formed by reasoning, calculation, and judgment to avoid numerous pointless trials and errors by restricting the number of possible choices in the conscious workspace. The consequences are great in mathematics, linguistics, esthetics, and ethics. One can conceive of the selection and memorization of an effective rule as the selection of a pattern of connections and the cooperation of neuronal assemblies, transmitted epigenetically within a social group and using a mechanism of imitation or shared reward. This is obviously of prime importance for anyone who, like me, is interested in the relationship between neuroscience and human science. This brings us back to the theme of Plato's three questions about the good, the true, and the beautiful, which has been our guideline throughout this book.

Enriching Knowledge

We first considered beauty through artwork and neuroesthetics. An artwork is destined to convey emotions between individuals. It possesses the evocative power to make conscious long-term unconscious memories and their emotional signatures, and to share them by empathy. So it has much significance. In his *Consciousness Regained* in 1984, Nicholas Humphrey referred to abstract art as "a visual poem built on rhyme and contrast between visual elements." At the neurobiological level, esthetic pleasure is supposed to mobilize in a concerted manner assemblies of neurons to bring together highly synthetic mental representations elaborated by the prefrontal cortex and the limbic system. The percept that, according to Onfray, fills us with astonishment and admiration by its esthetic efficacy corresponds, in my opinion, to a global access to multiple neural representations, current and remembered, visual and emotional, in the conscious workspace. The neural approach to esthetic experience is first and foremost through the visual system, the multiple, parallel, hierarchical pathways from the retina to the highest levels of the cerebral cortex. The visual system analyzes movement, depth, and spatial organization through its dorsal division, and it recognizes objects, faces, and colors through its ventral. Next come the emotional circuits in the limbic system, with subsections for motivation and desire (dopamine), aggression and anger (acetylcholine), fear and panic, and distress (opiates). The limbic system is intimately related to the prefrontal cortex. Recent fMRI studies by Thomas Jacobsen and colleagues (2006) demonstrated activation of the frontal, prefrontal, and posterior cingulate cortex for esthetic judgment, as opposed to parietal and premotor activity for judgment of symmetry (Figures 43, 44).

Epigenetic rules concerned in artistic creation include:

> *Correspondence with reality, or mimesis:* this rule of imitation by the artist is far from universally applied.
>
> *Consensus partium,* or harmony of the parts with the whole. This seems to me generally applied. Even in abstract works or pop art, there is composition.
>
> *Parsimony* is also widespread. Herbert Simon saw it as a criterion of beauty in a scientific theory, and of

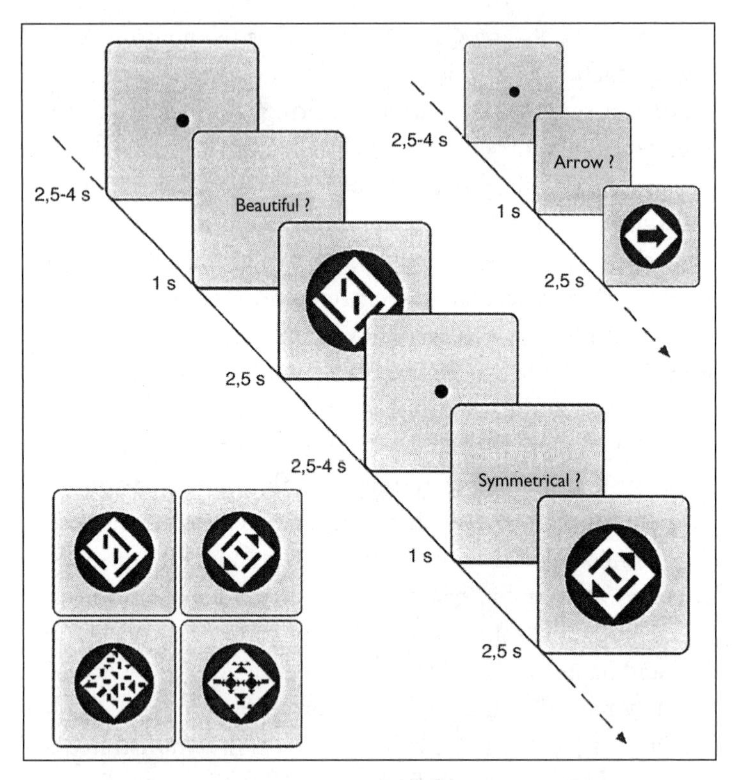

Figure 43: *A test of taste.* The subject has to decide if a figure is beautiful or simply symmetrical. From T. Jacobsen et al., "Brain Correlates of Aesthetic Judgment of Beauty," *NeuroImage* 32 (2006): 276–285; reprinted with permission from Elsevier.

course in an artwork insofar as it expresses much from little.

Novelty is universal. It combats esthetic fatigue and déjà-vu (in the sense of what we have already seen and what we have seen too much) and incites the artist to innovate.

Tranquility: the right to dream freely; catharsis.

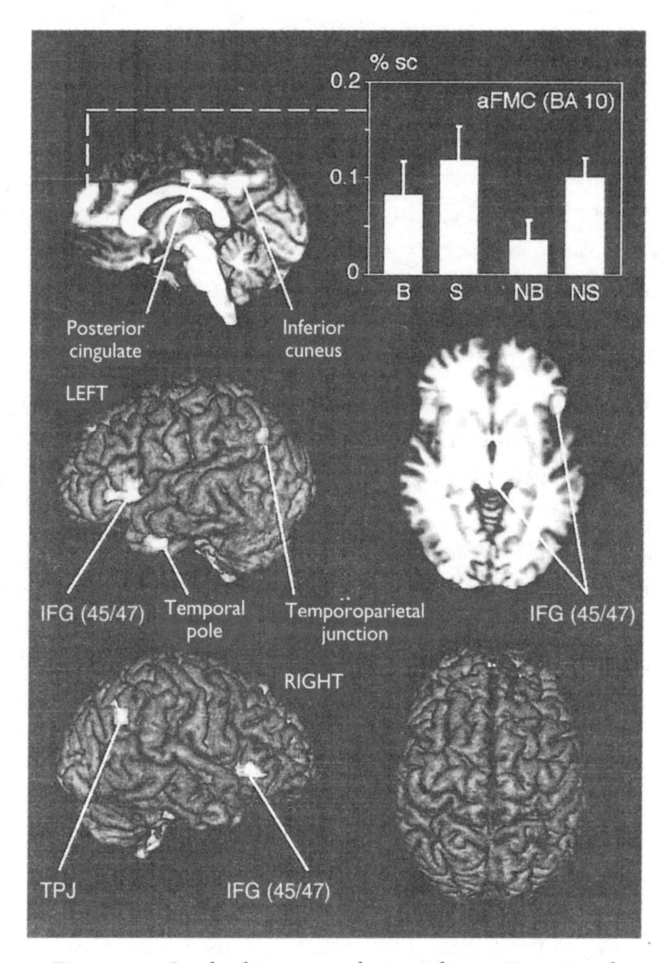

Figure 44: *Cerebral imagery of a test of taste.* Functional MRI images of the activated cortical areas for beauty or symmetry. From T. Jacobsen et al., "Brain Correlates of Aesthetic Judgment of Beauty," *NeuroImage* 32 (2006): 276–285; reprinted with permission from Elsevier.

> *The exemplum:* the idea that the artist conveys of his conception of the world and his universal ethical message.

Now let us consider esthetic values, already found in birds. Their evolutionary origins are multiple. Darwin emphasized *sexual selection:* beauty serves as publicity, as a signal of good health and evolutionary fitness. In Wilson and Sober's *group selection,* art serves to reinforce cooperative behavior for the overall survival of the species. *Art without imposed constraints reconciles man with himself* by reliving common esthetic experiences, at the intercommunal and also at the planetary level.

Moral norms, neuroethics, differ radically from the rules of art in that the former relate not to interpersonal communication of some esthetic pleasure, but to rules for living adopted deliberately and consciously by an individual within a social group, with a primary aim of resolving personal conflicts. Whereas art does not progress but is constantly renewed, moral norms evolve as lifestyle is improved by science and technology. Ethics committees, for example, have contributed to this evolution.

Neural elements involved in ethical norms include:

> *Empathy,* the capacity of the individual to comprehend the mental state of others, to credit them with knowledge, beliefs, and emotions, and to recognize differences and similarities between one's own mental states and theirs.
>
> *Inhibition of violence,* and *sympathy:* to avoid acts that make others suffer and, if needed, help them.

Epigenetic rules that contribute to the emergence of moral norms include the distinction between conventional prescriptions specific to a given culture and moral obligations common to most if not all societies. The Declaration of the

Rights of Man of 1789 and, above all, the Universal Declaration of Human Rights of 1948 express this universality.

There remain scientific representations. Their aim is to contribute to research for universal and cumulative objective truth, leading to progress in knowledge, technology, and industry. The neural predispositions include those already mentioned in terms of esthetics and ethics, with, in addition, the capacity for *distancing*, the differentiation between a self-centered, or egocentric, point of view, and one as seen from the outside, or allocentric. In this context epigenetic rules will above all be those of rationality: any proposition must obey the rules of logic. But critical examination and public debate are also important in order to select a solution that works, that is closest to reality and the most acceptable to all—and therefore universal. One of the critical features of science is to accept revision of established "truth," and equally to progress constantly toward understanding the world and ourselves. The dynamics of science depend just as much on the proposal of theoretical models, mental representations of an object or a process or a universe, as on public experimental proof, accompanied by validation, invalidation, evocation of new theories, imitation, innovation, and competition within an international network of scientists evolving in time and space.

The theme of the Good comes back to ethical norms and the quest for an individual happy life with others in a social group, surrounded by rational norms and the hope of an improved lifestyle. The theme of Truth is the aim of scientific research in pursuit of rational objective truths, universal and cumulative, with constant critical review but with progress in knowledge. The theme of Beauty brings together esthetic activities through interpersonal communication, involving emotion in harmony with reason, and reinforcing social bonds, not progressing but constantly renewing.

VII
Epilogue

The Significance of Death Seen by an Evolutionary Neurobiologist

Recent progress in neuroscience and its integration in dynamic evolutionary processes, which include culture and its history, prompt us to rethink certain central philosophical questions, such as the significance of death. Death is an essential biological phenomenon directly related to the evolution of species. It has taken on a special dimension in the history of humanity. Buffon rightly said that "death is as natural as life." Many philosophical and religious fundamentals, which emphasize the sacred character of life, maintain the balance by doing the same for its interruption by death. I feel it is opportune today more than ever to take up Buffon's concept and enliven it with what we have learned since about human cognition. In particular, we know that the human brain contains the neuronal basis for consciousness of self, which participates in every individual's quest to know about his unique destiny, and especially his death, with reference to memories of the past and intentions for the future. We also know that cognitive systems are especially well developed in humans; they intervene in social life, notably in the recognition of others as variants of ourselves, to whom we attribute the knowledge, beliefs, intentions, and emotions of a lifetime, which death inexorably terminates.

Hypotheses about the meaning that man has given to the phenomenon of death have evolved during the cultural history of mankind. "Sacred obscurity," to use the theologian Jacques Bénigne Bossuet's term, the impossible objective comprehension of death and the rejection of its being part of the natural order of things, has been propitious for the invention by human society of the most diverse myths. These myths have gradually given way to rational and scientific interpretations, immediately more consoling because they in no way consider the inevitability of death as evil or unavoidable suffering, but regard it as a simple fact of biological evolution.

DEATH IS NOT CHARACTERISTIC OF MAN ALONE

Man shares death with all higher life, animal and plant. Only micro-organisms like bacteria can escape. In the right conditions, their single cell divides into two identical cells that themselves can produce two more identical cells, and so on. Bacteria do in fact have sexual features, but these basic forms of life do not have a real mortal "body." If the environment becomes unfavorable, cell division stops. The cell can survive, or it may disappear by lysis. It can also form a resistant spore that reverts to a dividing cell when conditions again become favorable. Each cell is thus capable of surviving integrally in its own line of descent. It is a concrete form of eternal life, in no way mythical.

In evolution, death appears only with the multicellular organisms, such as sponges, jellyfish, corals, and *Hydra*. These organisms are composed of hundreds, even millions, of cells that differentiate into specific types within the organism, such as skin, muscle, or nerve, as well as germ cells for sexual reproduction. In 1883 August Weismann proposed two lines of cells. Germ cells form spermatozoa and oocytes, which multiply in the adult body and perpetuate life outside the body by fertilization. Somatic cells form the mortal body and have a life expectancy of from a few days to a century or more.

Perpetuation of life by sexual reproduction and fertilization involves a large loss of living material. This waste of energy is seen first in the organization of a body with a limited lifespan determined by the species' particular genetic makeup. This waste is even more obvious in the gametes, the male or female sex cells, released from the organism and only a small fraction of which actually produce offspring. In man most of these gametes, living cells if ever there were, die without participating in reproduction, either in wasted seminal emissions in the male or in menstrual flow in the female. We cannot really agree with either the legislator or the theologian that fertilization is the start of life. They may be haploid (containing half the number of chromosomes of the somatic cell), but spermatozoa and oocytes are definitely living cells, perhaps more fundamentally so than the diploid (twice the haploid number) zygote that they form by fusing. Indeed, the gametes form the basic link between generations. Even if their chances of survival are minute, the participation of a happy few is essentially analogous, without being identical, to the contribution of the single bacterial cell for the perpetuation of the species. So we can speak, as Weismann did, of the continuity of the germ plasm though the intermediary of gametes or, rather metaphorically, of an immortal germ line.

Even if only higher organisms possess a mortal body, this mortality is certainly not unique to man, who has inherited it through his evolutionary tree from far distant ancestors, probably from before sponges and coelenterates. The human genome contains specific determinants shared with its curious ancestors. These genes intervene during embryonic development, when the germ and somatic cell lines separate, to determine the life span of the body specific to the species. The first category of these determinants includes the numerous developmental genes (for example, the homeotic genes that determine the multiple, successive, parallel stages of gene expression associated with embryonic morphogenesis and cell differentiation). All these genes have been identified for *C. elegans,* in which the whole genome has been published. We find genes coding for transcription factors. Others code for molecules engaged in intercellular interaction and communication, as well as signal transduction, such as adhesion molecules, peptides, and receptors. It is from these that genes that define sex and the germ line are recruited, notably genes for programmed cell death (apoptosis) of certain well-defined embryonic cells. They are, for example, genes coding for caspases, proteolytic enzymes that when activated cause apoptosis. A second category of gene, less well defined, contributes to the temporal evolution of the body, its maturation and aging, and its death. We could mention genes for precocious aging in Werner syndrome, or those responsible for early-onset familial dementia of the Alzheimer type, like presenilin-1, amyloid precursor protein, or apolipoprotein E. Genes have recently been identified in *C. elegans* and *Drosophila* that can affect longevity: daf-12 mutants in *C. elegans* live three or four times longer than normal. Homologues of these genes could exist in man.

Besides these intrinsic genetic mechanisms, which delimit an envelope of longevity, extrinsic influences on lifestyle, such as medical progress, play an important role in slowing aging and prolonging life. In fifty years, life expectancy for women has increased by twenty years and in men by eighteen. But increasing the lifespan by improving self-imposed conditions does not alter the inevitability of death. Often the triggering factor is not fatal in itself, but leads to an irreversible decline. This could be myocardial infarction, a cerebrovascular accident, cancer, or infection. Advances in biomedical research mean that the trigger factors become easier to control every day, and we might expect that the concept of "natural" death will gradually disappear. Mastery of triggering circumstances may well soon remove the randomness of the moment of death in favor of a medical or personal decision on a predictable and deliberate ending to life. Will we in years to come expect medicalized death to become dignified euthanasia carried out with adequate palliative care?

The evolutionary significance of death is not related simply to the development of sex, which already existed in unicellular organisms and ensured genetic diversity that favored, in François Jacob's words, evolutionary "tinkering." The separation of the somatic from the germ line accompanied multicellularity, but what is its evolutionary advantage? The advent of multicellularity interests evolutionary theorists. The evolutionary transition that drove single cells to assemble into a complex organism could be understood, according to Richard Michod, on the basis of group selection, of cooperation between cells. Such an organism has a number of evolutionary advantages. First, the simple fact of cooperation between cells renders it less vulnerable to possible degradations of its various parts. Multicellularity also permits functional diversity within an organism and an increase in structural complexity and physiological specialization inaccessible to the single cell. Selection acts on the mortal phenotype, so having multiple specialized organs for nutrition, reproduction, perception, motor activity, cognition, and consciousness enormously increases the chances of survival. The body takes the germ line on board, thus protecting it from deleterious effects.

THE HUMAN BRAIN AND CONSCIOUSNESS OF DEATH

Still more important, our nervous system, including our brain, has developed throughout evolution like a cognitive prosthesis to improve the survival of our body. In the end we possess a brain simply because we are multicellular with a mortal body. More specifically, the human brain owes its origins to the evolutionary history of the higher mammals during the last four million years. Judging from its complexity, it has the metaphorical value of an organism within the organism. This "organ of the soul" is specialized in regulation, not only within the body but between human beings.

The brain is an integral part of the body. Nevertheless, its physiology is very special because its organization not only incorporates the genetic evolution of the species, which draws up the rough outline of its functional architecture. It also incorporates the epigenetic evolution of synaptic connections between neurons in the first stages of development, and in adulthood. Interaction with the outside world, physical, social, or cultural, leaves traces in neural networks by selectively stabilizing connections with those synapses that are effective. The result is memory and recall of information which may be stabilized even for a whole lifetime.

A consequence of this capacity for memory is the development of epigenetically transmissible culture from brain to brain within a social group. These tracks are generally labeled with emotional overtones, which is the signature of their selection. All representations produced by the brain result from selective mobilization of different populations of neurons by coherent chemical and electrical activity. The dynamic geography of these states of activity defines their meaning. These mental objects possess such an obvious material basis that it seems inconceivable to imagine their perpetuation—immortality—outside the mortal neuronal organization that produced them, and their epigenetic transmission. These mental objects are perceptive, motor, concrete, or abstract.

The human brain also possesses special systems for the representation of self as an immediate experience. Shaun Gallagher and Chris Frith recently proposed a neurocognitive model that differentiates the sense of self that causes an action, and involved the prefrontal and premotor cortex, from the sense of self that is subject to a sensory experience, which involves the cerebellum. Ricoeur and Gallagher also distinguished a narrative self, extending in time and including memories of the past and intentions for the future. Its tracks are probably widely distributed in the brain. Antonio Damasio recently attempted to overcome the difficulties that exist in determining the relationships between *core self* and *autobiographical self* by proposing an original neural model of this relationship. Even if still only partly developed, these ideas illustrate the plausibility of a neurobiology of self and its perception. This capacity to represent self, or rather several selves, is reinforced in infancy and later in adulthood by the capacity to represent others. This is found very early, initially by imitation, then by the attribution to others of mental states such as knowledge, beliefs, intentions, or emotions. Giacomo Rizzolatti's mirror neurons in the premotor cortex would be involved, and neuronal groups in the ventromedial prefrontal cortex would play an essential role in what one may call empathy and social reasoning.

The human brain thus possesses the ability to be aware of its own limited life span and to compare it with that of others. The emotional anchoring of the representation of self, especially autobiographical self, and of the self of others—close as well as distant—is, as Damasio showed, particularly powerful. This emotional component, combined with the stability of memory within the networks of our brains, disrupts the state of our brains when we face death. It leads to moral suffering. We can see in the suffering caused by the irreversible loss of someone else and, by anticipation, our own death, the brutal onset of a state of deprivation. There is an irremediable absence of a gratifying response, a violent disharmony between the expected and the real. The certainty of death, our own and that of others, seems absurd, unreasonable, and unjust because, outside the world of science, it does not offer any immediately comprehensible justification for its irremediable universality. Suffering in the face of death is suffering in the face of an inexorable destiny, the evolutionary causality of which seems unassailable, beyond both reason and will.

THE MYTHS OF DEATH

Homo sapiens buries his dead. He was the first in man's evolutionary lineage to have realized the tragic side of this biological phenomenon. His brain immediately sought causes or models or explanatory theories. The human brain's projective capacity appears very early. For instance, babies attribute intentions to simple geometrical figures moving on a computer screen. From an early age the human brain is an overproducer of meaning. It is remarkable that the first evidence of writing in China, on oracle bones of the Shang era, are related to divination; they try to give meaning to randomly occurring fracture lines in bone fragments that can have no real meaning.

Through his brain, man is both a rational and a social species. As Emile Durkheim wrote, the only means we have to free ourselves from physical force is to oppose it with collective force. Socially united human brains have thus produced collective representations as explanatory models that are effective on the social plane. They include such things as myths, beliefs, magic, and supernatural forces, and they are transmitted from generation to generation, from brain to brain. These collective representations allow us to organize time, not only our own but that observed objectively by all members of a given civilization. They bring interior peace and comfort by stimulating in our minds reward systems that involve transmitters or modulators such as dopamine and opiates. Dare we say that the opiate of the people takes on a neural meaning?

In his analysis of the archeology of religion, Colin Renfrew, in the footsteps of the anthropologist Roy Rappaport, emphasized the importance of the invention of mysterious supernatural agents, beyond human control, in these first explanatory models. Their invention offered a coherent vision of the nature and origins of the world, and of the future of humanity in situations in which, for primitive man, very few objective data were available. These representations, these mental objects, reconciled the fears of the individual with the life of the social group within which he found his reasons for survival. As these representations were at the level of the social group, they increased confidence and reinforced its cohesion. The imagery, and imagination, of these mythical representations even provided mnemotechnical support for their epigenetic transmission. "Symbolic entrainment" also played an essential role in the application of moral guidelines: divine punishment was symbolically more potent in the human mind than simple social sanctions.

So the significance of death resulted in numerous mythical representations throughout the diverse cultures that developed during recent human evolution. The commonest myth in our Western culture, sometimes termed Orphico-Platonic, is based on the dualism of body and soul. The activity of the brain is supposed to be separate from the mortal body and capable of autonomous, nondegradable perpetuation after death. This mythology, adopted by Judeo-Christianity, postulates the existence of a world beyond the grave where souls survive for eternity. Anguish, however, is not totally excluded, for there is no return from this hypothetical kingdom of the dead, and souls are subjected to a last judgment. In Eastern traditions the opposite is held, that life and death depend on immanent powers of nature, and one avoids addressing specific divinities with normative attributes outside human reach. Closer to our own culture, pre-Socratic Greek philosophers such as Democritus and Epicurus rejected dualism and saw death as a simple degradation of the biological organism. So one should not aspire to impossible immortality but accept the laws of nature with serenity, even if they are tragic. These Greek philosophers anticipated by 2,500 years the views of contemporary neuroscience. Today there is no objective scientific evidence for the immortal soul. All cerebral function stops when the EEG becomes flat. The activity of neuronal oscillators engaged in sleep-wake cycles ceases. Thalamocortical pathways fail. There is irreversible loss of consciousness. The brain ultimately decomposes and regulatory proteins disintegrate.

In spite of this scientific evidence, belief in the immortal soul and the resurrection of the dead persists, among other remarkable inventions of the human brain. For Durkheim this served an essential social function in contributing to ensuring perpetual life for the group and explaining the continuity of collective life. It is understandable why this belief is still rooted in our brains. This said, we are living at a time in social history when these tenacious myths, however powerful, are tending to lose their influence. There are limits to how much irrationality we will accept. We can imagine new collective representations whose symbolic power and possible secular ritualization may bring more comfort than the perspective of a pitiless last judgment forced on us by our natural biological destiny, our material decomposition, which none of us can avoid. Fyodor Dostoyevsky in *The Brothers Karamazov* put these words into the mouth of Ivan Fyodorovitch: "If there is no immortality of the soul, then there is no virtue, and everything is lawful." Pyotr Alexandrovitch replies that his theory is a fraud and states: "Humanity will find in itself the power to live for virtue even without believing in immortality. It will find it in love for freedom, for equality, for fraternity." We too should construct a collective ritual in solidarity to memory, an empathic accompaniment for the end of life that would make the inevitability of death tolerable for ourselves and others.

Conclusion

The Good, the True, and the Beautiful

In the third part of his *Elements of Physiology,* which deals with the brain, Diderot humorously suggested that "the wise man is only a compound of molecules," and a few lines later, "organization and life: that is the soul; only, the organization is so variable." My aim in this book has been to give a free hand to ideas about the molecule and about the soul, from bottom up as well as from top down, in the context of our brain. It has been to try to grasp, step by step, the place of the rather erratic evolution of this prodigious organ. *Tesseras agitans*—shaking the dice, as Diderot said.

We might envisage a program for the future in which there is no mention of imaginary "preestablished harmony," but a salutary asceticism that allows us to reflect on a new conception of our species, its origin and its future, on the basis of true multidisciplinary integration uniting neuroscience, biology, human science, and the history of civilization. This new approach leads us to reexamine Plato's three fundamental questions on the Good, the True, and the Beautiful, after abandoning their original essentialist context for a more unitary concept. All three imply epigenetic social representations, yet each is different.

But what is the meaning or use of all this? To help answer this question, I might mention a text by the Nobel Peace laureate René Cassin, "La science et les droits de l'homme" (1972), in which he emphasized the immense contribution of science to the conception, development, and practical respect of human rights. He recognized that industrial inventions and advances in rational medicine, from ancient Greece to modern biotechnology, had directly contributed to easing human suffering. Since the Renaissance, man's aspiration for liberty, notably of belief, has paralleled his desire for liberty of expression, an intrinsic part of the creative thinking of science. Scientific progress has indirectly favored the recognition of human rights. Today the breathtaking developments in neuroscience allow us to take a further step forward. As I said in *The Physiology of Truth,* better knowledge of humans and humanity enables us to "value the multiplicity of personal experience, the richness of cultural diversity, and the variety of our ideas about the world." This knowledge must "favor tolerance and mutual respect on the basis of a recognition of others as persons like ourselves, members of a single social species that is the product of biological evolution." Nevertheless, bearing in mind how our brains are made up, it will not happen spontaneously and effortlessly: it will always be a tough responsibility. In our fragile world, with its uncertain future, we must relentlessly drive the human brain to invent a future that will allow humanity to achieve a richer life with more solidarity with those around us.

Selected Bibliography

Aboitiz, F., and R. García. "The Evolutionary Origin of the Language Areas in the Human Brain. A Neuroanatomical Perspective." *Brain Research Reviews,* 1997 25 381–396.

Arendt, D., and K. Nübler-Jung. "Comparison of Early Nerve Cord Development in Insects and Vertebrates." *Development,* 1999 126 2309–2325.

Aristotle. *Metaphysics.* Translated by W. D. Ross. Oxford: Clarendon Press, 1924.

———. *Politics.* Translated by B. Jowett. New York: Dover Publications, 2000.

Atlan, H. *Entre le cristal et la fumée: Essai sur l'organisation du vivant.* Paris: Seuil, 2001.

Axelrod, R. *The Evolution of Cooperation.* New York: Basic Books, 1984.

Baars, B. J. *A Cognitive Theory of Consciousness.* New York: Cambridge University Press, 1988.

Baron-Cohen, S., A. Leslie, and U. Frith. "Mechanical, Behavioural and Intentional Understanding of Picture Stories in Autistic Children." *British Journal of Developmental Psychology,* 1986 4 113–125.

Barresi, J., and C. Moore. "Intentional Relations and Social Understanding." *Behavioral and Brain Sciences,* 1996 19 107–122.

Bernard, C. *Leçons sur les anesthésiques et sur l'asphyxie.* Paris: Baillière, 1874.

Berthoz, A. *Le sens du mouvement.* Paris: Odile Jacob, 1997.

———. *La décision.* Paris: Odile Jacob, 2003.

Bianchi, L. *La mécanique du cerveau et la fonction des lobes frontaux.* Paris: Librairie Louis Arnette, 1921.

Bloomfield, L. *Language.* New York: Henry Holt, 1933.

Bourdieu. P. *Pascalian Meditations* (1997). Translated by R. Nice. Stanford: Stanford University Press, 2000.

Bourgeois, J.-P., and P. Rakic. "Changes of Synaptic Density in the Primary Visual Cortex of the Macaque Monkey from Fetal to Adult Stage." *Journal of Neuroscience,* 1993 13 2801–2820.

Bourgeois, J.-P., A. Ryter, A. Menez, P. Fromageot, P. Boquet, and J.-P. Changeux. "Localization of the Cholinergic Receptor Protein in *Electrophorus electroplax* by High Resolution Autoradiography." *FEBS Letters,* 1972 25 27–133.

Bourgeois, L. *Solidarité.* Lormont: Bord de l'Eau, 2008.

Buser, P., and M. Imbert. *Neurophysiologie fonctionnelle.* Paris: Hermann, 1975.

Canguilhem, G. *Du développement à l'évolution au XIXe siècle.* Paris: PUF, 1962.

Cartaud, J., E. L. Benedetti, J. B. Cohen, J. C. Meunier, and J.-P. Changeux. "Presence of a Lattice Structure in Membrane Fragments Rich in Nicotinic Receptor Protein from the Electric Organ of *Torpedo marmorata.*" *FEBS Letters,* 1973 33 109–113.

Cassin, R. "La science et les droits de l'homme." *Impact: Science et Société,* 1972 22 359–369.

Cavalli-Sforza, L. *Gènes, peuples et langues.* Paris: Odile Jacob, 1996.

———. *Evolution biologique, évolution culturelle.* Paris: Odile Jacob, 2005.

Changeux, J.-P. "The Feedback Control Mechanisms of Biosynthetic L-Threonine Deaminase by L-Isoleucine." *Cold Spring Harbor Symposia on Quantitative Biology,* 1961 26 313–318.

———. "Sur les propriétés allostériques de la L-thréonine désaminase de biosynthèse. VI. Discussion générale." *Bulletin de la Société de Chimie Biologique,* 1965 47 281–300.

———. "Remarks on the Symmetry and Cooperative Properties of Biological Membranes." In A. Engström and B. Strandberg, eds., *Symmetry and Function of Biological Systems at the Macromolecular Level. Nobel Symposium 11.* New York: Wiley Interscience, 1969, 235–256.

———. *Neuronal Man: The Biology of Mind.* Translated by Laurence Garey. New York: Pantheon Books, 1985.

———. *The Physiology of Truth.* Translated by M. B. DeBevoise. Cambridge: Harvard University Press, 2004.

Changeux, J.-P., and A. Connes. *Conversations on Mind, Matter, and Mathematics.* Princeton: Princeton University Press, 1995.
Changeux, J.-P., P. Courrège, and A. Danchin. "A Theory of the Epigenesis of Neural Networks by Selective Stabilization of Synapses." *Proceedings of the National Academy of Sciences,* 1973 70 2974–2978.
Changeux, J.-P., and S. Dehaene. "Neuronal Models of Cognitive Functions." *Cognition,* 1989 33 63–109.
———. "Hierarchical Neuronal Modeling of Cognitive Functions: From Synaptic Transmission to the Tower of London." *Comptes Rendus de l'Académie des Sciences,* 1998 321 241–247.
Changeux, J.-P., and S. J. Edelstein. *Nicotinic Acetylcholine Receptors.* Paris: Odile Jacob, 2005.
Changeux, J.-P., M. Kasai, and C. Y. Lee. "The Use of a Snake Venom Toxin to Characterize the Cholinergic Receptor Protein." *Proceedings of the National Academy of Sciences,* 1970 67 1241–1247.
Changeux, J.-P., and A. Taly. "Nicotinic Receptors, Allosteric Proteins and Medicine." *Trends in Molecular Medicine,* 2008 14 93–102.
Cheney, D. L., and R. M. Seyfarth. *How Monkeys See the World.* Chicago: University of Chicago Press, 1990.
Chomsky, N., *Aspects of the Theory of Syntax.* Cambridge: MIT Press, 1965.
———. "Language and Nature." *Mind,* 1995 104 1–61.
Crick, F. *The Astonishing Hypothesis.* New York: Scribner's, 1994.
Crick, F. and C. Koch. "Some Reflections on Visual Awareness." *Cold Spring Harbor Symposia on Quantitative Biology,* 1990 55 953–962.
Damasio, A. R. "The Somatic Marker Hypothesis and the Possible Functions of the Prefrontal Cortex." *Philosophical Transactions of the Royal Society London, Biological Sciences,* 1996 351 1413–1420.
Darwin, C. *On the Origin of Species.* London: John Murray, 1859.
———. *The Descent of Man and Selection in Relation to Sex.* London: John Murray, 1871.
———. *The Expression of the Emotions in Man and Animals.* London: John Murray, 1872.
Dawkins, R. *The Selfish Gene.* Oxford: Oxford University Press, 1976.
Deacon, T. *The Symbolic Species.* New York: W. W. Norton, 1997.
Dehaene, S., and J.-P. Changeux. "A Simple Model of Prefrontal Cortex Function in Delayed-Response Tasks." *Journal of Cognitive Neuroscience,* 1989 1 244–261.
———. "The Wisconsin Card Sorting Test: Theoretical Analysis and

Simulation of a Reasoning Task in a Model Neuronal Network." *Cerebral Cortex,* 1991 1 62–79.

Dehaene, S., J.-P. Changeux, L. Naccache, J. Sackur, and C. Sergent. "Conscious, Preconscious, and Subliminal Processing: A Testable Taxonomy." *Trends in Cognitive Sciences,* 2006 10 204–211.

Dehaene, S., M. Kerszberg, and J.-P. Changeux. "A Neuronal Model of a Global Workspace in Effortful Cognitive Tasks." *Proceedings of the National Academy of Sciences,* 1998 95 14529–14534.

Dejerine, J. "Contribution à l'étude anatomo-pathologique et cliniques des différentes variétés de cécité verbale." *Mémoires de la Société de Biologie,* 1892 4 61–90.

———. *Anatomie des centres nerveux.* Paris: Rueff, 1895.

Denton, D. *The Primordial Emotions: The Dawning of Consciousness.* New York: Oxford University Press, 2005.

Descartes, R. *Dioptrics* (1637). Translated by J. Cottingham, R. Stoothoff, and D. Murdoch. Cambridge: Cambridge University Press, 1984.

———. *The Passions of the Soul* (1649). Translated by J. Cottingham, R. Stoothoff, and D. Murdoch. Cambridge: Cambridge University Press, 1984.

———. *Treatise on Man* (1664). Translated by J. Cottingham, R. Stoothoff, and D. Murdoch. Cambridge: Cambridge University Press, 1984.

De Waal, F. *Good Natured: The Origins of Right and Wrong in Humans and Other Animals.* Cambridge: Harvard University Press, 1996.

Dissanayake, E. *Homo aestheticus: Where Art Comes From and Why.* New York: Free Press, 1992.

Dubois, J., L. Hertz-Pannier, G. Dehaene-Lambertz, Y. Cointepas, and D. Le Bihan. "Assessment of the Early Organization and Maturation of Infants' Cerebral White Matter Fiber Bundles: A Feasibility Study Using Quantitative Diffusion Tensor Imaging and Tractography." *NeuroImage,* 2006 30 1121–1132.

Edelman, G. M. *Neural Darwinism: The Theory of Neuronal Group Selection.* New York: Basic Books, 1987.

———. *Topobiology: An Introduction to Molecular Embryology.* New York: Basic Books, 1988.

———. *The Remembered Present: A Biological Theory of Consciousness.* New York: Basic Books, 1989.

Edelman, G. M., and G. Tononi. *A Universe of Consciousness: How Matter Becomes Imagination.* New York: Basic Books, 2000.

Elston, G. N. "Cortex, Cognition and the Cell: New Insights into the

Pyramidal Neuron and Prefrontal Function." *Cerebral Cortex,* 2003 13 1124–1138.

Fishman, Y. I., I. O. Volkov, M. C. Noh, P. C. Garell, H. Bakken, J. C. Arezzo, M. A. Howard, and M. Steinschneider. "Consonance and Dissonance of Musical Chords: Neural Correlates in Auditory Cortex of Monkeys and Humans." *Journal of Neurophysiology,* 2001 86 2761–2788.

Fodor, J. *The Language of Thought.* New York: Crowell, 1975.

Fontenelle, B. de. *Conversations on the Plurality of Worlds* (1686). Translated by E. Gunning. London: T. Hurst, 1803.

Fransson, P., B. Skiöld, S. Horsch, A. Nordell, M. Blennow, H. Lagercrantz, and U. Aden. "Resting-State Networks in the Infant Brain." *Proceedings of the National Academy of Sciences,* 2007 104 15531–15536.

Freud, S. *Project for a Scientific Psychology* (1895). In *Complete Psychological Works of Sigmund Freud.* 24 vols. London: Hogarth Press, 1953, 1:283–397.

Gehring, W. J. *Master Control Genes in Development and Evolution: The Homeobox Story.* New Haven: Yale University Press, 1998.

Gisiger, T., M. Kerszberg, and J.-P. Changeux. "Acquisition and Performance of Delayed-Response Tasks: A Neural Network Model." *Cerebral Cortex,* 2005 15 489–506.

Gould, S. J., and R. C. Lewontin. "The Spandrels of San Marco and the Panglossian Paradigm: A Critique of the Adaptationist Programme." *Proceedings of the Royal Society of London, Biological Sciences,* 1979 205 581–598.

Grice, H. P. "Meaning." *Philosophical Review,* 1957 66 377–388.

Hacking, I. "The Looping Effect of Human Kinds." In D. Sperber, D. Premack, and A. Premack, eds., *Causal Cognition: An Interdisciplinary Approach.* Oxford: Clarendon Press, 1995, 351–383.

Hamilton, W. D. "The Evolution of Altruistic Behavior." *American Naturalist,* 1963 97 354–356.

Heidmann, T., and J.-P. Changeux. "Fast Kinetic Studies on the Interaction of a Fluorescent Agonist with the Membrane-Bound Acetylcholine Receptor from *T. marmorata.*" *European Journal of Biochemistry,* 1979 94 255–279.

Hein, G., and T. Singer. "I Feel How You Feel but Not Always: The Empathic Brain and Its Modulation." *Current Opinion in Neurobiology,* 2008 18 153–158.

Holloway, R. "Toward a Synthetic Theory of Human Brain Evolution." In J.-P. Changeux and J. Chavaillon, eds., *Origins of the Human Brain*. Oxford: Oxford University Press, 1995, 42–60.

Jackendoff, R. *Consciousness and the Computational Mind*. Cambridge: MIT Press, 1987.

Jacob, F., and J. Monod. "Genetic Regulatory Mechanisms in the Synthesis of Proteins." *Journal of Molecular Biology*, 1961 3 318–356.

———. "On the Regulation of Gene Activity." *Cold Spring Harbor Symposia on Quantitative Biology*, 1961 26 193–209.

Jacobsen, T., R. I. Schubotz, L. Höfel, and D. Y. Cramon. "Brain Correlates of Aesthetic Judgment of Beauty." *NeuroImage*, 2006 29 276–285.

James, W. *The Principles of Psychology*. New York: Henry Holt, 1890.

Jouvet, M. *The Paradox of Sleep: The Story of Dreaming*. Cambridge: MIT Press, 1999.

Kandel, E. R., J. H. Schwartz, and T. M. Jessell. *Principles of Neural Science*. 4th ed. New York: McGraw-Hill, 2000.

Kerszberg, M., and J.-P. Changeux. "A Model for Reading Morphogenetic Gradients: Autocatalysis and Competition at the Gene Level." *Proceedings of the National Academy of Sciences*, 1994 91 5823–5827.

———. "A Simple Molecular Model of Neurulation." *BioEssays*, 1998 20 758–770.

Kropotkin, P. *Mutual Aid: A Factor of Evolution* (1902). Boston: Extending Horizons Books, 1955.

Lamarck, J. B. *Zoological Philosophy* (1809). Translated by H. Elliot. London: Macmillan, 1914.

Land, E. H. "Experiments in Color Vision." *Scientific American*, 1959 200 84–99.

Laureys, S. "The Neural Correlate of (Un)wareness: Lessons from the Vegetative State." *Trends in Cognitive Sciences*, 2005 9 556–559.

Le Brun, Charles. *L'expression des passions et autres conférence et correspondance*. Edited by J. Philipe. Paris: Dédale, 1994.

Lévi-Strauss, C. *Race and History*. Paris: UNESCO, 1952.

Li, G. D., D. C. Chiara, G. W. Sawyer, S. S. Husain, R. W. Olsen, and J. B. Cohen. "Identification of a $GABA_A$ Receptor Anesthetic Binding Site at Subunit Interfaces by Photolabeling with an Etomidate Analog. *Journal of Neuroscience*, 2006 26 11599–11605.

Livingstone, M. *Vision and Art: The Biology of Seeing*. New York: Harry N. Abrams, 2002.

Livingstone, M. S., and D. H. Hubel. "Effects of Sleep and Arousal on the Processing of Visual Information in the Cat." *Nature,* 1981 291 554–561.

Llinás, R. R., and D. Paré. "Of Dreaming and Wakefulness. *Neuroscience,* 1991 44 521–535.

Locke, J. *Essay concerning Human Understanding.* London: Edward Mory, 1690.

Marr, D. *Vision.* San Francisco: W. H. Freeman, 1982.

Monod, J., J.-P. Changeux, and F. Jacob. "Allosteric Proteins and Cellular Control Systems." *Journal of Molecular Biology,* 1963 6 306–329.

Monod, J., J. Wyman, and J.-P. Changeux. "On the Nature of Allosteric Transitions: A Plausible Model." *Journal of Molecular Biology,* 1965 12 88–118.

Onfray, M. *Archéologie du présent: Manifeste pour une esthétique cynique.* Paris: Grasset, 2003.

Paré, D., and R. R. Llinàs. "Conscious and Preconscious Processes as Seen from the Standpoint of Sleep-Walking Cycle Neurophysiology." *Neuropsychologia,* 1995 33 1155–1168.

Pavlov, I. P. *Conditioned Reflexes* (1927). Translated by G. V. Anrep. New York: Dover Publications, 2003.

Peirce, C. S. *Collected Papers.* Edited by C. Hartshorne and P. Weiss. 8 vols. Cambridge: Harvard University Press, 1931–1960.

Penrose, R. *The Emperor's New Mind.* Oxford: Oxford University Press, 1989.

Plato. *Dialogues: Hippias Major, Republic, Symposium.* Translated by B. Jowett. New York: Random House, 1920.

Popper, K. R. *The Logic of Scientific Discovery* (1934). London: Hutchinson, 1959.

———. *Conjectures and Refutations: The Growth of Scientific Knowledge.* London: Routledge and Keegan Paul, 1963.

———. *Objective Knowledge: An Evolutionary Approach.* Oxford: Clarendon Press, 1972.

Premack, D., and A. Premack. "Intention as Psychological Cause." In D. Sperber, D. Premack, and A. Premack, eds., *Causal Cognition.* Oxford: Clarendon Press, 1995, 185–200.

Premack, D., and G. Woodruff. "Does the Chimpanzee Have a Theory of Mind?" *Behavioral and Brain Sciences,* 1978 4 515–526.

Prins, H. H. T. *Ecology and Behaviour of the African Buffalo: Social Inequality and Decision Making.* London: Chapman and Hall, 1996.

Ribary, U. "Dynamics of Thalamo-Cortical Network Oscillations and Human Perception." *Progress in Brain Research,* 2005 150 124–142.

Ricoeur, P. *Oneself as Another* (1990). Translated by K. Blarney. Chicago: University of Chicago Press, 1992.

Sabouraud, O. *Le langage et ses maux.* Paris: Odile Jacob, 1995.

Saussure, F de. *Cours de linguistique générale.* Paris: Payot, 1979.

Searle, J. "Consciousness." *Annual Review of Neuroscience,* 2000 23 557–578.

Shallice, T. *From Neuropsychology to Mental Structure.* Cambridge: Cambridge University Press, 1988.

Skinner, B. F. *Verbal Behavior.* New York: Appleton-Century-Crofts, 1957.

Spencer, H. *Social Statics.* London: Chapman, 1851.

———. *First Principles.* London: Williams and Norgate, 1862.

Sperber, D., and D. Wilson. *Relevance: Communication and Cognition.* Oxford: Blackwell, 1986.

Spinoza, B. de. *The Ethics* (1677). London: Bell, 1883.

Sukhotinsky, I., V. Zalkind, J. Lu, D. A. Hopkins, C. B. Saper, and M. Devor. "Neural Pathways Associated with Loss of Consciousness Caused by Intracerebral Microinjection of GABA$_A$-Active Anesthetics." *European Journal of Neuroscience,* 2007 25 1417–1436.

Turing, A. M. "Computing Machinery and Intelligence." *Mind,* 1950 49 433–460.

———. "The Chemical Basis of Morphogenesis." *Philosophical Transactions of the Royal Society London, Biological Sciences,* 1952 237 37–72.

Unwin, N. "Nicotinic Acetylcholine Receptor and the Structural Basis of Fast Synaptic Transmission." Croonian Lecture 2000. *Philosophical Transactions of the Royal Society London, Biological Sciences,* 2000 355 1813–1829.

Venter, J. C., et al. "The Sequence of the Human Genome." *Science,* 2001 291 1304–1351.

Vygotsky, L. *Thought and Language* (1934). Translated by E. Hanfmann and G. Vakar. Cambridge: MIT Press, 1962.

Watson, J. B. *Behaviorism.* New York: People's Institute Publishing, 1925.

Wilson, D. S., and E. Sober. "Reintroducing Group Selection to the Human Behavioral Sciences." *Behavioral and Brain Sciences,* 1994 17 585–654.

Young, J. Z. *A Model of the Brain.* Oxford: Oxford University Press, 1964.

Zeki, S. *Inner Vision: An Exploration of Art and the Brain.* Oxford: Oxford University Press, 1999.
Zigmond, M. J., F. E. Bloom, S. C. Landis, J. L. Roberts, and L. R. Squire. *Fundamental Neuroscience.* San Diego: Academic Press, 1999.

Index